Diet for Iron Deficiency

Klaus Günther

Diet for Iron Deficiency

Metabolism - Bioavailability - Diagnostics

Klaus Günther
Institut für Bio- und Geowissenschaften und Institut für Ernährungs- und Lebensmittelwissenschaften
Forschungszentrum Jülich und Universität Bonn
Jülich und Bonn, Nordrhein-Westfalen, Germany

ISBN 978-3-662-65607-5   ISBN 978-3-662-65608-2   (eBook)
https://doi.org/10.1007/978-3-662-65608-2

This Springer imprint is published by the registered company Springer-Verlag GmbH, DE, part of Springer Nature.
The registered company address is: Heidelberger Platz 3, 14197 Berlin, Germany

*For my 3 dear stars
and all companions
on this fantastic journey,
called life.*

Preface

Iron deficiency is the most common deficiency disease in the world, even though a whopping 5% of the element is present in over 100 different minerals in the continental crust, mostly as oxide, oxide hydroxide, sulfide, or carbonate. This has always left me in a strong state of astonishment, and I have wondered what reasons could possibly be responsible. Within the primary nutritional goals of the World Health Organization (WHO), one of the top priorities in the near future since the 2012 World Health Assembly (WHA) in Geneva is the global reduction of anemia, most of which is caused by iron deficiency. Reason enough, therefore, for chemists, biochemists, and food scientists to concern themselves intensively with this goal and the topic.

Much has happened in iron biochemistry recently, and about 20 years ago hepcidin entered the scene, a central regulator of iron balance whose function was previously unknown. Numerous iron enzymes have now been precisely characterized and their function understood in detail, including in brain metabolism. The mechanisms of iron homeostasis in the organism have been described in more detail, and the properties of ferritin in food and its resorption in the organism now appear in a new light.

In this reference book, I would like to approach the complex subject from the food chemistry and biochemistry side, where I myself have many years of research experience, but of course not lose sight of the overall picture. Therefore, new nutritional aspects and diagnostic topics from clinical chemistry and laboratory medicine will also be discussed here.

Since my doctoral thesis at the University of Münster and the habilitation period at the University of Bonn in the 1990s, I have been concerned with questions of the bioavailability of toxic and essential elements in food and their significance for humans. Within my various research groups, the clarification of the bonding forms of the elements was a particular focus of interest. The entire periodic table was always in focus, since the individual elements and of course their species influence each other during absorption in the human organism. The declared aim was a multi-element species analysis in order to obtain the most complete possible insight into the binding relationships in food.

In order to work on this interdisciplinary topic, it was necessary to master an extensive spectrum of methods, ranging from the selective determination of the various elements in the trace range to powerful separation methods and the modern structure elucidation of complexing agents and proteins by high-resolution mass spectrometry.

Furthermore, I have been working at the Research Centre Jülich for many years. The outstanding and forward-looking science managers at the center have understood very well how to further orient this large research institute with over 7000 employees towards the pressing issues of the future, which of course also include improving the global food situation.

The bioeconomy is now one of the three focal areas, alongside research activities in the fields of energy and information. Within the bioeconomy, securing the world's food supply with high-quality food plays a central role, and plant-based foods in particular are increasingly coming into focus, as their production can be far more climate friendly than is possible with animal products. The raw materials and nutrition turnarounds, climate protection, and the energy turnaround are global challenges that affect all areas of life.

These topics are also being given increasing consideration in my lectures at the University of Bonn, where they are meeting with an excellent response. Due to clever and sustainable strategies of the governing bodies, a motivating spirit of optimism in many areas of the alma mater and also by taking into account these new interdisciplinary, global contexts, the University of Bonn was able to win six clusters of excellence in the federal government's excellence strategy, more than any other university in Germany, and has been a university of excellence since 2019.

For example, the joint excellence cluster "PhenoRob" of the Faculty of Agriculture at the University of Bonn and the Research Centre Jülich is investigating the potential uses of robotics for sustainable crop production. This is an important new research unit that will also improve the production of high-quality plant-based foods with a high content of essential minerals, trace elements, and other micronutrients in the future. These can then be used against the "hidden hunger" in the world, including the "hidden iron hunger."

Because through cleverly composed, high-quality foods, you can powerfully counteract an iron deficiency and avoid it. This is what I would like to show you in this book. There are new and exciting scientific findings here, which I will discuss in detail.

Furthermore, it is important that you do not fall into the iron trap. Due to the predominant consumption of quite healthy foods, which unfortunately contain very little iron, this can happen quickly. This happens more often than you think, especially with nutrition-conscious people, a hitherto underestimated point of view, which I describe in the book with examples, and which surprised me quite a bit.

I have garnished these central messages with many new scientific findings from iron biochemistry and the bioavailability of the element from food, which of course cannot be complete, as they would go beyond the scope of the book. Furthermore, there are many references to new publications on the subject, which invite professional deepening of the material.

Simple rules for iron supply with natural foods must not be missing even in a reference book. Therefore, the most important facts are summarized in concise rules of thumb at the end of the book.

In the event of a suspected iron deficiency or illness, this book is of course no substitute for a visit to the doctor, and at no point are recommendations for therapy made. However, it does show simple ways in which one can significantly improve one's iron status through the targeted selection of natural foods.

Plant iron is much more valuable than previously thought. This is a central message of this book.

For reasons of better readability, the generic masculine is predominantly used in this book. This always implies both forms, thus including the female form.

Bonn, Germany   Klaus Günther
October 2022

Contents

About the Author

 **Klaus Günther** works at the Research Centre Jülich, a member of the Helmholtz Association, Germany's largest scientific organization, and teaches as a professor at the Institute of Nutrition and Food Sciences at the Rhenish Friedrich Wilhelm University of Bonn. In his research work, the biochemist and food scientist has for many years been concerned with the bioavailability of minerals and trace elements in food and their significance for humans. He is the author of numerous scientific publications in international journals and has served as a member of various expert committees, including those of federal authorities. Professor Günther headed C4/W3 chairs at the Universities of Duisburg-Essen and Bonn for many years. He was appointed honorary professor at the Technical Erzherzog Johann University in Graz, Austria, and at the State Key Laboratory of Food Science and Technology at Nanchang University, P.R. China.

Introduction

1

Within the world's population, 25% are affected by severe iron deficiency, which in most cases is the cause of anemia. This is over 1.5 billion people. If we now add to this the people who have a milder form of iron deficiency without anemia and in whom important bodily functions and enzyme activities are thus already impaired, the number of people affected increases even more considerably. This makes iron deficiency the most significant deficiency worldwide. This fact highlights the magnitude of what is at stake, and the great interest in this topic is reflected in recent review articles addressing the current state of etiology, pathophysiology, epidemiology, diagnostics and therapy (Cappellini et al. 2020; Lopez et al. 2016).

Eliminating iron deficiency is a very important task in the context of activities to improve world nutrition, as well as the overall global supply of micronutrients (hidden hunger) (Willett et al. 2019; Haddad et al. 2016). For example, one of the global goals of the World Health Organization (WHO) is to reduce anemia in women of reproductive age by 50% by 2025 (WHA 2012; WHO 2014). A major issue, then, with political dimensions, of course. It is also interesting in this context that the iron content of important food crops will decrease significantly in the coming years with the expected increase in CO_2 in the atmosphere, thus exacerbating the problem (Myers et al. 2014).

This is also the context of the global discussion on increasing the iron content of plant-based foods, which can then be used to prevent iron deficiency in the diets of the world's population (Vasconcelos et al. 2017). This so-called biofortification means overall the enrichment of important nutrients (e.g. also zinc, calcium, essential fatty and amino acids, antioxidants and vitamins) in plant-based foods through classical breeding measures or even green genetic engineering. In 2016, for example, the World Food Prize was awarded for the topic of biofortification (enrichment of iron, zinc and beta-carotene in vegetables or cereals).

An iron deficiency is the cause of 80% of anemia. Through specifically selected food combinations, you can strongly counteract an iron deficiency, even with

K. Günther, *Diet for Iron Deficiency*,
https://doi.org/10.1007/978-3-662-65608-2_1

vegetarian and vegan lifestyles or other special diets. This is what I would like to show you in this book and present you with simple solutions. Furthermore, I will present new findings about the biochemistry and bioavailability of iron and outline current developments in the diagnosis of iron deficiency.

Blood consists approximately half of liquid blood plasma and the other half of various cells. Erythrocytes make up a large proportion of these cells (Fig. 1.1). These transport oxygen from the lungs to the organs and tissues. The red blood cells develop from stem cells of the bone marrow. This process is called erythropoiesis. The availability of iron is also very important in this process.

Erythrocytes were first described in 1658 by the physician Jan Swammerdam (1637–1680) from Amsterdam. He was one of the first researchers to use the microscope, which was significantly developed by the Dutchman Antoni van Leeuwenhoek (1632–1723) and which was an important prerequisite for the beginning of scientific hematology (Fig. 1.2).

After the microscopic discovery of red blood cells, the history of iron in the blood has been very chequered and again required the interaction of numerous scientific disciplines (Coley 2001; Beutler 2002; Poskitt 2003). In 1713 – a good 50 years after the first description of erythrocytes – iron was first discovered in blood by the French chemists and physicians Nicolas Lemery (1645–1715) and Francois Geoffroy (1672–1731) (see Fig. 1.3). The blood was carbonized by them. This process produced magnetic iron particles, which they were able to detect with the aid of a magnet.

It then took another 100 years for these findings to be generally accepted and for the Swedish biochemist Jöns Jakob Berzelius to finally confirm them, as it was not

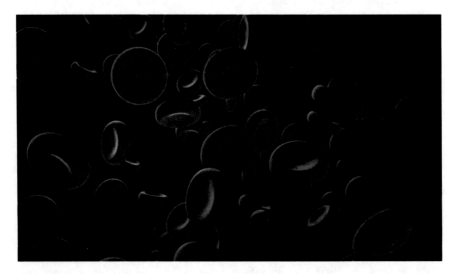

Fig. 1.1 Representation of erythrocytes. They consist of 90% of the dry mass of hemoglobin and have the shape of biconcave discs with mean diameters of 7.5 μm. The thickness is 2 μm at the edge and 1 μm in the middle in humans. Average life span 4 months, development time 7 days, new production about 1%/day = about 200 billion/day or 2 million/s. (© flashmovie/stock.adobe.com)

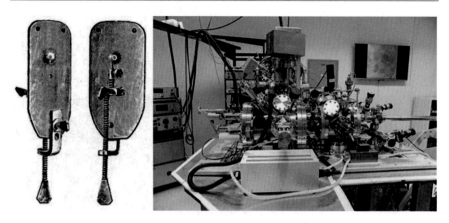

Fig. 1.2 Model of a Leeuwenhoek microscope (left, essential part is a small glass bead in the round recess) and modern electron microscope (right). The development of microscopic techniques is not over yet. In 2017, for example, the Nobel Prize in Chemistry was awarded for the development of cryoelectron microscopy, which can significantly increase the resolution when studying biomolecules. The high-performance electron microscope "Pico" is operated at Forschungszentrum Jülich, with a resolving power of 50 pm, i.e. much smaller than the atomic distance in chemical bonds (C–C single bond length: 154 pm). The pm range is 1 million times smaller than the micrometer range of erythrocytes (Fig. 1.1). (left: © Juulijs/stock.adobe.com, right: © Martina/stock.adobe.com)

Fig. 1.3 The College de France in Paris, the place of work of Francois Geoffroy, who was also very well known for his tables listing chemical affinities, in which iron (fer, French) already had its place. (© David Bostock/stock.adobe.com)

Fig. 1.4 Stamp with the portrait of Justus von Liebig (1803–1873). He worked as a chemistry professor in Giessen and Munich and was significantly involved in the scientific development of analytical chemistry and food chemistry. On the right you can see his five-ball apparatus for the determination of carbon, a symbol of the beginning of exact, quantitative chemistry. His meat extract, a nutrient concentrate that was very well known in the nineteenth century, is shown on the left. (© konstantant/stock.adobe.com)

until the beginning of the nineteenth century that the knowledge to carry out precise analyses was first possessed. This art was developed in particular in the famous experimental school of Justus von Liebig in Giessen (Fig. 1.4).

A further 100 years later, detailed investigations of the iron content of the blood began with the discovery of the element in the red blood cells and the description of the various iron-binding forms hemoglobin, myoglobin, iron enzymes, ferritin, hemosiderin and transferrin, the structure and function of which are described in the following Chap. 2. The essential work on this took place in the twentieth century.

Then, at the beginning of the twenty-first century, the view of the biochemistry and physiology of iron was expanded once again. Summary book publications on the element's effect on mental processes, brain function, and the pathology of iron deficiency and excess, diagnosis and clinical aspects of central nervous system diseases, and public health issues reflect this now even more interconnected view (Yehuda and Mostofsky 2010).

The discovery of the central role of hepcidin in the control of iron balance (Nicolas et al. 2001) and the elucidation of the molecular processes involved in the regulation of erythropoiesis and the oxygen uptake of cells by the hypoxia-induced factor HIF also gave new and important impetus to research into iron biochemistry. In 2019, the Nobel Prize in Medicine was awarded to the Briton Ratcliffe and the two US-Americans Kaelin and Semenza for their discoveries in connection with HIF (cf. Sect. 3.1).

As can be seen, research on the subject has been in a constant state of flux since the discovery of iron in the blood in 1713, and even today many questions surrounding iron biochemistry in the body are still unanswered, and we are only just beginning to really understand the extensive functions that this element has to fulfil in the human body, in addition to its well-known roles in oxygen transport.

In nutrition and food science, too, traditional assessments are currently being put to the test. For example, plant iron is much more valuable than previously thought, and new uptake pathways for ferritin iron in intestinal cells are being discussed, which cast the bioavailability of phyto-iron in a new light. The reference values for the daily intake of iron of the international professional societies are currently the subject of intense discussion, as they differ in part considerably, and scientific studies on the forms of binding of iron in the various foods are more topical than ever, as they have a significant influence on the absorption in the organism.

In summary, the topic of nutrition and iron is a highly interesting and exciting field of activity for the various scientific disciplines, where surprising findings can be obtained time and again, especially at their interfaces.

References

Beutler E (2002) History of iron in medicine. Blood Cells Mol Dis 29:297–308

Cappellini MD, Musallam KM, Taher AT (2020) Iron deficiency anaemia revisited. J Intern Med 287:153–170

Coley NG (2001) Early blood chemistry in Britain and France. Clin Chem 47:2166–2178

Haddad L, Hawkes C, Waage J, Webb P, Godfray C, Toulmin C (2016) Food systems and diets: facing the challenges of the 21st century. Global Panel on Agriculture and Food Systems for Nutrition, University of London Institutional Repository

Lopez A, Cacoub P, Macdougall IC, Peyrin-Biroulet L (2016) Iron deficiency anaemia. Lancet 387:907–916

Myers S, Zanobetti A, Kloog I, Huybers P, Leakey A, Bloom A, Carlisle E, Dietterich L, Fitzgerald G, Hasegawa T, Holbrook N, Nelson R, Ottman M, Raboy V, Sakai H, Sartor K, Schwartz J, Seneweera S, Tausz M, Usui Y (2014) Rising CO_2 threatens human nutrition. Nature 510:139–142

Nicolas G, Bennoun M, Devaux L, Beaumont C, Grandchamp B, Kahn A, Vaulont S (2001) Lack of hepcidin gene expression and severe tissue iron overload in upstream stimulatory factor 2 (USF2) knockout mice. Proc Natl Acad Sci U S A 98:8780–8785

Poskitt EME (2003) Early history of iron deficiency. Br J Haematol 122:554–562

Vasconcelos MW, Gruissem W, Bhullar NK (2017) Iron biofortification in the 21st century: setting realistig targets, overcoming obstacles, and new strategies for healthy nutrition. Curr Opin Biotechnol 44:8–15

Willett W, Rockström J, Loken B, Springmann M, Lang T, Vermeulen S, Garnett T, Tilman D, DeClerck F, Wood A, Jonell M, Clark M, Gordon L, Fanzo J, Hawkes C, Zurayk R, Rivera J, De Vries W, Sibanda L, Afshin A, Chaudhary A, Herrero M, Agustina R, Branca F, Lartey A, Fan S, Crona B, Fox E, Bignet V, Troell M, Lindahl T, Singh S, Cornell S, Reddy S, Narain S, Nishtar S, Murray C (2019) Food in the anthropocene: the EAT-lancet commission on healthy diets from sustainable food systems. Lancet 393:447–492

World Health Assembly HA65.6 (2012) Comprehensive implementation plan on maternal, infant and young child nutrition. In: Sixty-fifth World Health Assembly Geneva, 21–26 May 2012. Resolutions and decisions, annexes. World Health Organization, Geneva; 2012:12–13. Accessed 6 Oct 2014

World Health Organization (2014) Global targets 2025. To improve maternal, infant and young child nutrition. Accessed 6 Oct 2014

Yehuda S, Mostofsky D (eds) (2010) Iron deficiency and overload: from basic biology to clinical medicine. Humana Press, Springer Science+Business Media LLC, New York

Biochemistry of Iron

2

Iron metabolism is still an intensive field of research today, and numerous questions remain to be answered. Nevertheless, it is possible to approach the subject clearly, and accordingly I would like to present a simple classification at the beginning.

Basically, there are five different main biochemical forms of iron around which the most important things in iron metabolism revolve. These are roughly summarized in Fig. 2.1.

In these various compounds, iron (configuration [Ar] $3d^6$ $4s^2$) can exist in two different oxidation forms: iron-II (Fe^{2+}, configuration [Ar] $3d^6$) and iron-III (Fe^{3+}, configuration [Ar] $3d^5$). Iron-II is also referred to as ferro and iron-III as ferri form. Depending on the nature of the ligands surrounding the iron ions, high-spin or low-spin complexes can form, in which the electrons are maximally unpaired or paired, respectively (Binnewies et al. 2016).

The chemistry of the two forms in solution differs considerably. In the presence of oxygen and in the absence of special ligands, the ferric ion is the more stable form. However, like Fe-II, it does not occur as a simple hydrated cation $[Fe(H_2O)_6]^{3+}$ at physiological pH (7.4) because it is much too acidic. Deprotonation of the aqua ligands to the OH group stabilizes the higher oxidation state of iron, and partially insoluble hydroxides with different water contents are formed at pH values near the neutral point.

The complicated interrelationships of redox potential and pH, taking into account the solubility products from the sparingly soluble phase, can be illustrated in Pourbaix diagrams, which are often referred to in discussions of the biochemistry of the element. Iron chemistry in the body is thus dependent on numerous auxiliary molecules which can compensate for these special chemical peculiarities of the element in the aqueous milieu under physiological conditions.

K. Günther, *Diet for Iron Deficiency*, https://doi.org/10.1007/978-3-662-65608-2_2

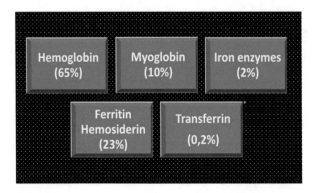

Fig. 2.1 Occurrence of iron in the human body and average rounded percentage distribution. By far the largest proportion of iron is found in hemoglobin and thus in erythrocytes. Although only 2% of iron is found in enzymes, this proportion is nevertheless extremely important. A large number of different iron enzymes are known, which are involved in the most diverse essential metabolic processes

2.1   The Heme Group

The heteroaromatic pyrrole with nitrogen as heteroatom is the basic building block of heme. When four substituted pyrrole rings with 4 methylene groups combine to form a ring, the tetrapyrrole system uroporphyrinogen-III or simply urogen-III is obtained (Fig. 2.2). It is the central biosynthetic precursor for the porphyrin system in heme, the corrin scaffold in vitamin B_{12} and the modified porphyrin in the green leaf pigment chlorophyll. In the finished organic molecules, iron (heme), cobalt (B_{12}) or magnesium (chlorophyll) are then complexed via the nitrogens of the pyrrole units.

Porphyrin is formed from urogen-III by dehydrogenation and modification of the side chains, resulting in the formation of a conjugated double bond system that is a Hückel aromatic with 18 π-electrons. After complexation of Fe-II, heme, a chelate complex, is then formed, with four coordination sites of the Fe-II ion occupied by the four nitrogen atoms of the pyrrole rings. Two of the pyrrole units have each donated a proton and are therefore present as anions, neutralizing the double positive charge of the Fe-II ion. The overall complex is therefore neutral, and the conjugated double bond system results in the four coordinative bonds of the nitrogen atoms being equivalent (Fig. 2.3).

In hemoglobin, the fifth coordination site is then occupied by the free electron pair of a histidine N atom and the sixth site can finally bind oxygen as an attachment complex, no oxidation of Fe-II occurs. However, if the Fe-II in heme is oxidized to Fe-III in the body by various processes, methemoglobin is formed. Although it can still bind oxygen, it can no longer release it. To prevent this, the body has its own redox systems. The elucidation of the structure of the porphyrin ring, which is very important for life, was achieved at the beginning of the twentieth century by the chemists Richard Willstätter (Nobel Prize 1915) and Hans Fischer (Nobel Prize 1930).

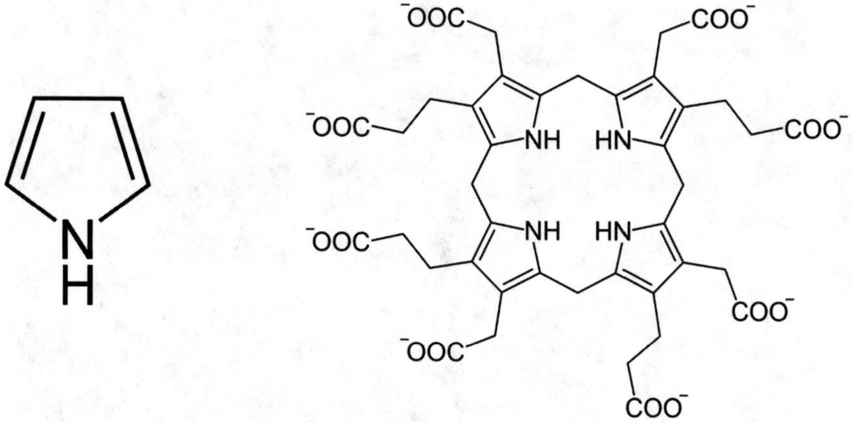

Fig. 2.2 Structures of pyrrole and urogen-III, the central biosynthetic precursor of heme

Fig. 2.3 Structure of the heme iron. The iron sits in the center of the porphyrin ring. The oxygen binds directly to the divalent iron (**Fe-II**), which is attached to the porphyrin skeleton via 4 nitrogen atoms (**N**). (© Leonid Andronov/ stock.adobe.com)

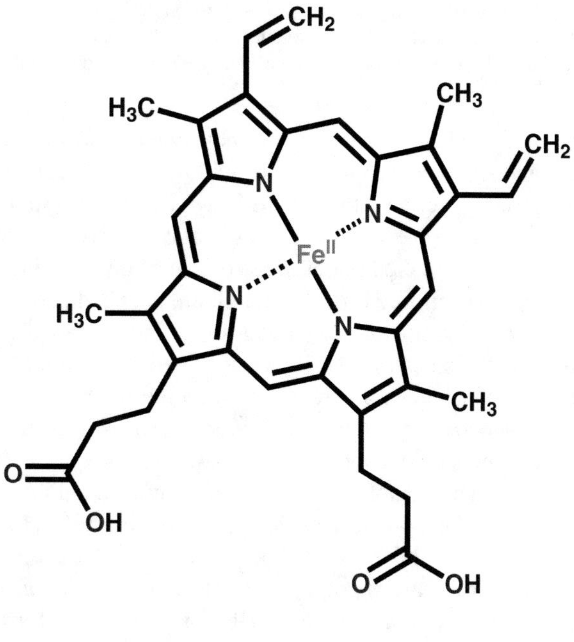

2.2 Myoglobin and Hemoglobin

The important functions of myoglobin and hemoglobin lie primarily in the storage and transport of oxygen. The heme iron is embedded in a protein matrix, which determines the special properties of the system for reversible oxygen binding to the iron central atom.

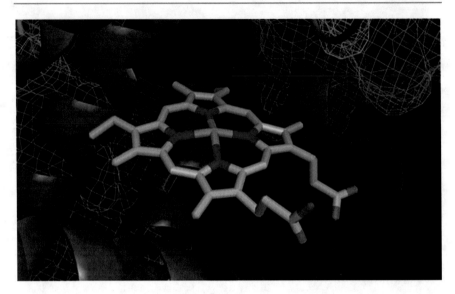

Fig. 2.4 Model for myoglobin. The heme-iron molecule (cf. Fig. 2.3, light red in the center iron, blue nitrogen) is packed in α-protein helices, which are connected by turns and form a spherical structure. Myoglobin is composed of a single polypeptide chain. (© Catalin/stock.adobe.com)

Myoglobin is responsible for the oxygen supply of the muscles (Greek *mys* muscle and Latin *globus* sphere) (Fig. 2.4). Myoglobin was the first protein ever for which an exact structure could be given after crystallization and X-ray structure analysis. John Kendrew and Max Perutz received the Nobel Prize in Chemistry in 1962 for the elucidation of the structure of myoglobin and hemoglobin.

The oxygen-free form of myoglobin is called deoxymyoglobin and the oxygen-containing compound oxymyoglobin. The so-called proximal histidine from the protein chain occupies another coordination site on the iron. A so-called distal histidine contributes a hydrogen bond to the binding of the oxygen molecule. In the case of deoxymyoglobin, one coordination site remains vacant and the iron lies slightly outside the porphyrin ring. After oxygenation, it shifts to the heme plane.

Human hemoglobin is composed of four myoglobin-like subunits (Fig. 2.5), two identical α- and β-chains each. This is very important for the special binding properties to oxygen and makes the molecule more effective. The four units cooperate with each other to ensure good uptake of oxygen in the lungs and extensive delivery to target organs. Hemoglobin A (HbA) is a tetramer consisting of a pair of identical α-β-dimers.

A hemoglobin can bind four oxygen molecules cooperatively, which means that the tendency for the fourth O_2 molecule to be taken up is much stronger than that for the first. This is caused by distinct changes in the quaternary structure during the attachment of oxygen. The quaternary structure present in deoxyhemoglobin is called the T-form (T = tense) and the structure in oxyhemoglobin is called the R-form (R = relaxed). The binding of oxygen causes a transition of the hemoglobin

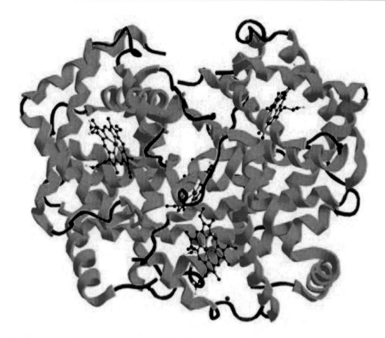

Fig. 2.5 Structure of hemoglobin, simplified it can be described as a fourfold myoglobin (four heme-irons in protein shells). The four units work together cooperatively in oxygen transport in the sense of molecular teamwork. (© raimund14/stock.adobe.com)

tetramer from the T- to the R-form. This increases the binding affinity of the other sites (Stryer et al. 2018).

Allosteric regulation of the strength of oxygen binding can also occur through hydrogen ions and carbon dioxide. This is called the Bohr effect, named after Christian Bohr, who first described this property of hemoglobin in 1904. Carbon monoxide has a much higher affinity for hemoglobin than oxygen, binds to the same site of iron as oxygen, and therefore prevents the formation of oxyhemoglobin and is therefore a respiratory toxin.

Due to the accumulation of oxygen, the hemoglobin changes its color. The arterial blood loaded with oxygen is scarlet, while the venous blood has a purple color. In addition, the magnetic properties are different in the two states because of the different electronic configurations. Oxyhemoglobin is diamagnetic and deoxyhemoglobin is paramagnetic.

The change in magnetic properties can be used in the examination of the brain by functional magnetic resonance imaging (fMRI). This imaging technique uses signals that come predominantly from the protons of water molecules that are altered by the change in magnetic properties of hemoglobin. A more active brain region contains a higher proportion of oxygenated hemoglobin. Thus, the relative activities of different brain areas can be distinguished.

Specific mutations in the hemoglobin subunit genes can cause disease. In sickle cell anemia, a single amino acid substitution in the β-chain of hemoglobin is responsible. A glutamic acid has been replaced by valine. The reason for the name of this disease is the abnormal sickle shape of erythrocytes in the absence of oxygen, which causes numerous physiological disadvantages for the body. The inherited disease thalassemia, or Mediterranean disease because it is more common in countries around the Mediterranean, is caused by an imbalanced production of α- and β-hemoglobin chains. In α-thalassemia, the α-chains of hemoglobin are not produced in sufficient amounts. This results in the formation of hemoglobin tetramers that contain only β-chains, bind oxygen with higher affinity but without cooperativity, and therefore little oxygen is released in the tissues. In β-thalassemia, not enough β-chains are formed, and the α-chain tetramers form insoluble aggregates that precipitate in mature erythrocytes (Stryer et al. 2018).

Vegetable Leghemoglobin
Very interesting in the context of nutrition is leghemoglobin, which was first isolated from legumes, hence the name, a long time ago (Kubo 1939). It is a heme protein found in micromolar amounts in roots of legumes infected by nodule bacteria (rhizobia). This important symbiosis allows nitrogen fixation, in which the nitrogenase complex plays an important role, being oxygen sensitive. The leghemoglobin has a dual function due to its high affinity for oxygen. On the one hand, it protects the nitrogenase, and on the other hand, it supplies the respiratory chain in the system by transporting oxygen, where the energy for the fixation of nitrogen is generated. In the active center of the nitrogenase complex is a cluster of iron, molybdenum and sulfur.

Leghemoglobin consists of the heme group and a single protein chain and is thus more comparable to myoglobin or the alpha or beta subunits of hemoglobin. The apoprotein is produced by plant cells the heme by bacteria. Leghemoglobin has a much higher affinity for oxygen than animal heme proteins, and the differences in electronic structure are also reflected in ^{57}Fe-Mößbauer spectra (Alenkina et al. 2018). Leghemoglobin is used in the recipe for meat substitutes and is supposed to give them the special characteristic taste (meat tasty).

2.3   Ferritin and Hemosiderin

Quantitatively also very important in the distribution of iron in the body are ferritin and hemosiderin. In both cases, this is depot iron, and the element is not present as heme iron but as iron hydroxide oxide micelles, i.e. in a compound with oxygen, in its trivalent form. About 4000 iron ions are clustered together in the middle of a shell consisting of 24 proteins (Fig. 2.6). By weight, ferritin consists of about 20% iron at normal saturation.

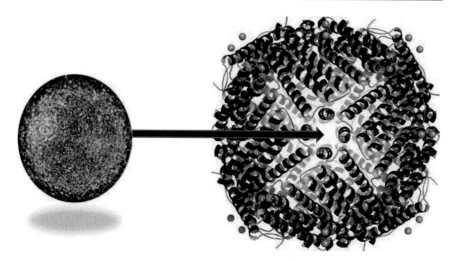

Fig. 2.6 Structure of ferritin. The iron core consists of about 4000 iron ions (grey sphere, left) and is located in the center of the sheath consisting of more than 20 different proteins (represented by helices). Individual grey spheres on the surface of the protein shell symbolize iron ions that can enter or leave the core through channels in the shell (iron storage function of the molecule). (Right image: © molekuul.be/stock.adobe.com)

Ferritins are found in humans, animals, plants and bacteria. They serve as storage molecules for iron, are water-soluble and can release iron when needed, e.g. for the production of hemoglobin. Ferritins are mainly located in the liver and bone marrow. The concentration of ferritin in the blood serum is a meaningful measure of the iron stores of the entire human organism (cf. Chap. 5).

The regulation for the expression level of ferritin takes place in the animal organism at the level of translation. The ferritin mRNA contains an IRE (iron response element) in the non-translated section. An IRP (IRE-binding protein) can bind to this structure, blocking the initiation of translation. As the iron concentration increases, IRP binds iron as a 4Fe–4S cluster and is released from the mRNA, and ferritin is formed.

In legumes, iron is largely present as ferritin and plays an important role in the newly discovered uptake pathway of plant iron. You will learn more about this and can thus participate in interesting new findings of recent years (cf. Chap. 7).

When there is a high supply of iron in the organism, the water-insoluble hemosiderin is formed by the accumulation of many ferritins. The iron components of the ferritins bind to each other with partial degradation of the proteins, form large particles, and therefore the iron content in the hemosiderin is much higher than in the ferritin.

2.4   Transferrin

Transferrin is a glycoprotein containing two ferric ions bound to specific binding sites (Fig. 2.7). It binds iron in the serum and transports it to the cells where it is taken up by transferrin receptors. The saturation of transferrin with iron plays an important role in determining iron deficiency in the organism (cf. Chap. 5).

Currently, the two different transferrin receptors 1 and 2 are known (TfR1 and TfR2), whereby TfR1 is synthesized in all cells, TfR2 mainly in the liver. Via receptor-mediated endocytosis, iron-loaded transferrin is taken up by the cell after binding to the receptor. In endosomes, the Fe-III ion dissolves from transferrin due to low pH, while apotransferrin remains bound to the receptor. The apotransferrin-receptor complex then moves to the plasma membrane, and the neutral environment of the extracellular fluid then dissociates the complex and the process can begin again at the cell surface. Disturbances in TfR1 or TfR2 biochemistry lead to the clinical picture of hematochromatosis, an iron overload of the organism.

Proteolytic detachment of TfR from the cell membrane results in soluble transferrin receptors (sTfR), which are freely present in the plasma and whose determination plays a role in the diagnosis of iron deficiency states (cf. Chap. 5). The concentration of sTfR in serum is directly proportional to the concentration of TfR on cell membranes. In the case of iron deficiency, the concentration of sTfR increases.

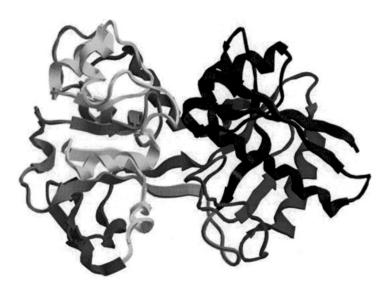

Fig. 2.7 Structure of transferrin. Only two individual iron-III ions are enclosed in the protein chains – and not an accumulation of several thousand as in the case of ferrritin – and then transported in this form to the sites in the body where iron is required. (© raimund14/stock.adobe.com)

2.5   Iron Enzymes

Enzymes are biocatalysts that make many reactions in organisms possible. If the function of enzymes is impaired, the associated metabolic pathways no longer run optimally and diseases can develop. Iron is a component of many enzymes and therefore, in addition to its function in the transport and storage of oxygen, also plays a very important role in many other central processes in the body.

So far, about 100 enzymes have been described in more detail in which iron is an essential factor. There are three types of iron proteins with heme groups: The oxygen-binding proteins myoglobin and hemoglobin described above, and the two enzyme families that activate oxygen molecules and are responsible for electron transport.

Oxygen activators include cytochrome oxidase, peroxidase, catalase and cytochrome P 450 s. They have a penta-coordinated geometry whereby the sixth position of the iron center can bind either molecular oxygen or hydrogen peroxide. In the case of cytochrome P 450 s, an iron-carbon bond forms with the substrate (Crichton 2016).

Cytochromes P450 (CYP) contain a heme iron thiolate group and play a central role in the metabolism of cholesterol and hormones, in detoxification mechanisms or in the degradation of drugs in the body (Fig. 2.8).

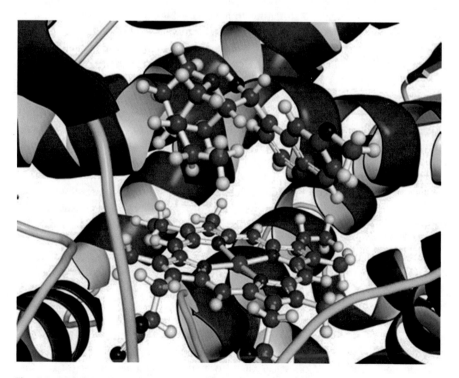

Fig. 2.8 Molecular view of a specific cytochrome (CYP2D6) with heme iron (lower molecule group with blue iron ion in the middle) and a drug to be degraded (upper molecule group: anti-malarial quinine). Both molecules are again embedded in a protein environment. The iron is additionally coordinated via a sulfur atom (thiolate) of a cysteine residue. CYPs are heme iron thiolate enzymes. (© molekuul.be/stock.adobe.com)

The cytochromes are a large group of monooxygenases that are responsible for biotransformation phase 1, which contain heme iron with a thiolate coordination and have their own designation system, e.g. CYP1A1, CYP1A2, CYP1B1, etc. In biotransformation phase 1, hydroxylations are carried out on molecules in the liver, thus functionalizing them for further conversions in phase 2, after which they can be excreted as water-soluble substances. The activities of the CYP enzymes can vary considerably between patients. Since they are also responsible for the degradation of drugs, this is very important for the dosage and is a current research topic in personalized medicine.

In the heme enzymes responsible for electron transport, iron is present as a hexacoordinated low-spin system. This involves the 3 enzyme classes cytochromes a, b and c, which are involved in electron transport during oxidative phosphorylation, among other things (Crichton 2016). The iron in these enzymes can switch between the Fe-III low-spin oxidation states with an unpaired electron and Fe-II low-spin. Because the iron remains in the low-spin state in both oxidation states, the landing switch is greatly facilitated. Low-spin and high-spin refer to the different occupation of the d orbitals of the iron central ion split in the ligand field in the concept of chemical ligand field theory (Binnewies et al. 2016).

In addition to the family of iron enzymes with a heme system, another separate class exists: the iron-sulfur enzymes. Their active sites are thought to be among the very first catalytic chemical structures to have found application in metabolic processes in nature (Crichton 2016; Huber and Wächtershäuser 1997) and are an interesting object of research in inorganic biochemistry, which has established itself as an independent field in recent years (Herres-Pawlis and Klüvers 2017; Kaim et al. 2013). The active sites of iron-sulfur enzymes also form biomimetic templates for new synthetic structural analogues and artificial enzymes (Lee et al. 2014).

Simple iron-sulfur enzymes can be broadly divided into two groups. Either iron alone is coordinated via sulfur from cysteine side chains (rubredoxins), which is a component of protein chains, or the direct bonds to iron occur through inorganic sulfide and the thiol groups of the side chains of cysteines. Complex iron-sulfur enzymes may additionally contain molybdenum and flavins.

In the case of simple iron-sulfur clusters, four basic structures exist, of which the most structurally complex, a cubane unit, is shown in Fig. 2.9. The cubane coordination structure as a whole can take three oxidation states $[Fe_4-S_4]^+$, $[Fe_4-S_4]^{2+}$ or $[Fe_4-S_4]^{3+}$ (Crichton 2016).

Enzymes with $[Fe_4-S_4]$ centers include, for example, aconitase (citrate cycle) or NADH dehydrogenase (respiratory chain), the structure $[Fe_2-S_2]$ is found in xanthine oxidase (nucleotide metabolism) or ferrochelatase (incorporation of iron into the porphyrin ring in heme) (see Fig. 2.3).

In addition to the described protein families with heme-iron and iron-sulfur clusters, iron as a simple coordinated cation functions as a cofactor in the following enzymes, among others, which will serve as examples here: Alcohol dehydrogenase (iron or zinc), linoleoyl-CoA desaturase (formation of arachidonic acid from linoleic acid, prostaglandin metabolism), ribonucleoside diphosphate reductase (pyrimidine metabolism), cysteine dioxygenase (taurine synthesis), phenylalanine

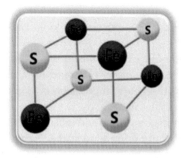

Fig. 2.9 Schematic structure of an iron sulfur center Fe_4 -S_4 of the Cuban type. Furthermore, Fe_2–S_2 – and other centers with different geometrical arrangements and bonding ratios also exist. The iron ions are further coordinated respectively with the thiol groups of four cysteines, which are located in protein chains. The chemical structure shown is also included in the logo of the research group of Richard Holm of the Chemistry Department of Harvard University, USA, which has done a great deal of work on bioinorganic iron-sulfur clusters. (See also Lee et al. 2014)

hydroxylase (formation of tyrosine from phenylalanine), tyrosine hydroxylase (catecholamine biosynthesis), tryptophan hydroxylase (serotonin and melatonin synthesis), beta-carotene 15,15′-monooxygenase (retinol metabolism). Some of these enzymes are described in more detail in Sect. 2.6.

All in all, this was only a small selection of enzymes in which iron fulfils an important function, and it thus becomes clear which complex health disadvantages can arise in the case of iron deficiency, which do not relate directly to the oxygen supply through myoglobin and hemoglobin. The 2% iron in the body, which is essential in enzymatic processes (cf. Fig. 2.1), thus has extremely diverse tasks in the entire metabolism of the organism. To further demonstrate this, the following two sections will take a closer look at the functions of iron in brain metabolism (Sect. 2.6) and in the immune system (Sect. 2.7) as examples.

2.6   Iron in Brain Metabolism

What is generally rather unknown: Iron is also extremely important for the function of the brain. It is involved in many core processes, such as the coating of sensitive nerves (myolinization), the formation of connections between nerve cells (dentritogenesis) or the production of important messenger substances (neurotransmitters).

Iron directly affects brain performance and even mental well-being. In adolescents and children, the risk of psychiatric disorders is increased with iron deficiency. These include mood swings, autistic spectrum disorders and ADHD (attention deficit hyperactivity disorder). This should once again illustrate how important this trace element is for the function of our thinking and control organs.

Iron therefore also plays a significant role in thought processes and even states of mind. Both of these are of course highly interesting in view of the problems that can be associated with them. Studies have shown that even in the case of an iron deficiency which is not yet characterized by anemia, i.e. which is not detected at all in a normal laboratory check only via the hemoglobin value, the emotional state of people can be disturbed.

I would like to pick out a few examples here and present them at the molecular level. The neurotransmitters dopamine, norepinephrine and epinephrine, among others, fulfil important functions in brain metabolism, and iron is essential in the synthesis of these catecholamines in the body. Dopamine has a predominantly excitatory effect on the central nervous system and is associated with increased drive and motivation and is often popularly referred to as the "happiness hormone."

The starting compound in the biosynthesis of catecholamines is tyrosine, which is produced from the amino acid phenylalanine under the catalytic influence of phenylalanine hydroxylase (PAH) by oxidation with oxygen. Fe-II is essential for the function of PAH (Fig. 2.10).

In the congenital hereditary disease phenylketonuria (PKU), this reaction of phenylalanine to tyrosine is disturbed. In children, this leads to maldevelopment of the brain with severe mental retardation. Phenylketones are then formed via other degradation pathways and excreted in the urine.

Phenylalanine hydroxylase (PAH), which is involved in this important reaction, contains an Fe^{2+} ion coordinated across different amino acid side chains. It is the enzyme of the AAAH family (AAAH: aromatic amino acid hydroxylase) that has been best characterized. This mechanism of hydroxylation is also found in other AAAHs, of which we will learn two more. The proposed reaction mechanisms in the active site of the PAH are shown in Fig. 2.11. The structure of the known cofactor BH4 can be found in Fig. 2.12.

The tyrosine formed in this biosynthesis is then hydroxylated to dopa by oxygen in a subsequent reaction. This reaction is also catalyzed by an Fe^{2+}-dependent enzyme belonging to the AAAM group, tyrosine hydroxylase, which has a very similar catalytic mechanism to phenylalanine hydroxylase (Fig. 2.13).

Dopa is then decarboxylated to form dopamine, a biogenic amine that acts predominantly as a neurotransmitter in the central nervous system. Dopamine is then converted into norepinephrine by the introduction of an OH group into the aliphatic side chain, and finally into epinephrine by methylation of the amino group. Both substances are very well known as important messengers in the brain and stress hormones, which are released during flight and fear stimuli.

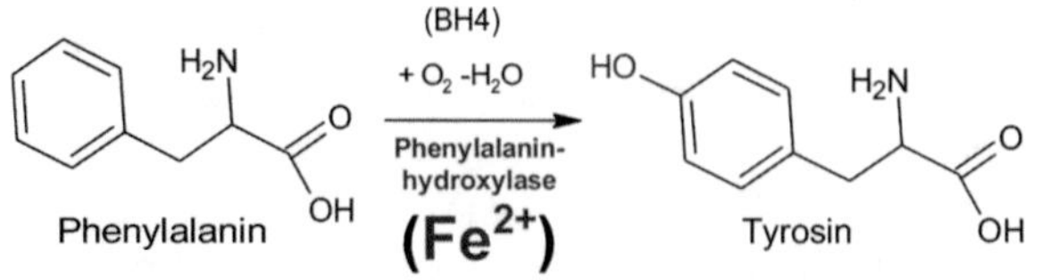

Fig. 2.10 Biosynthesis of tyrosine from phenylalanine under catalysis of the iron-containing phenylalanine hydroxylase. Oxidation occurs via oxygen using tetrahydrobiopterin (BH4) as a redox cofactor. (Waløen et al. 2017)

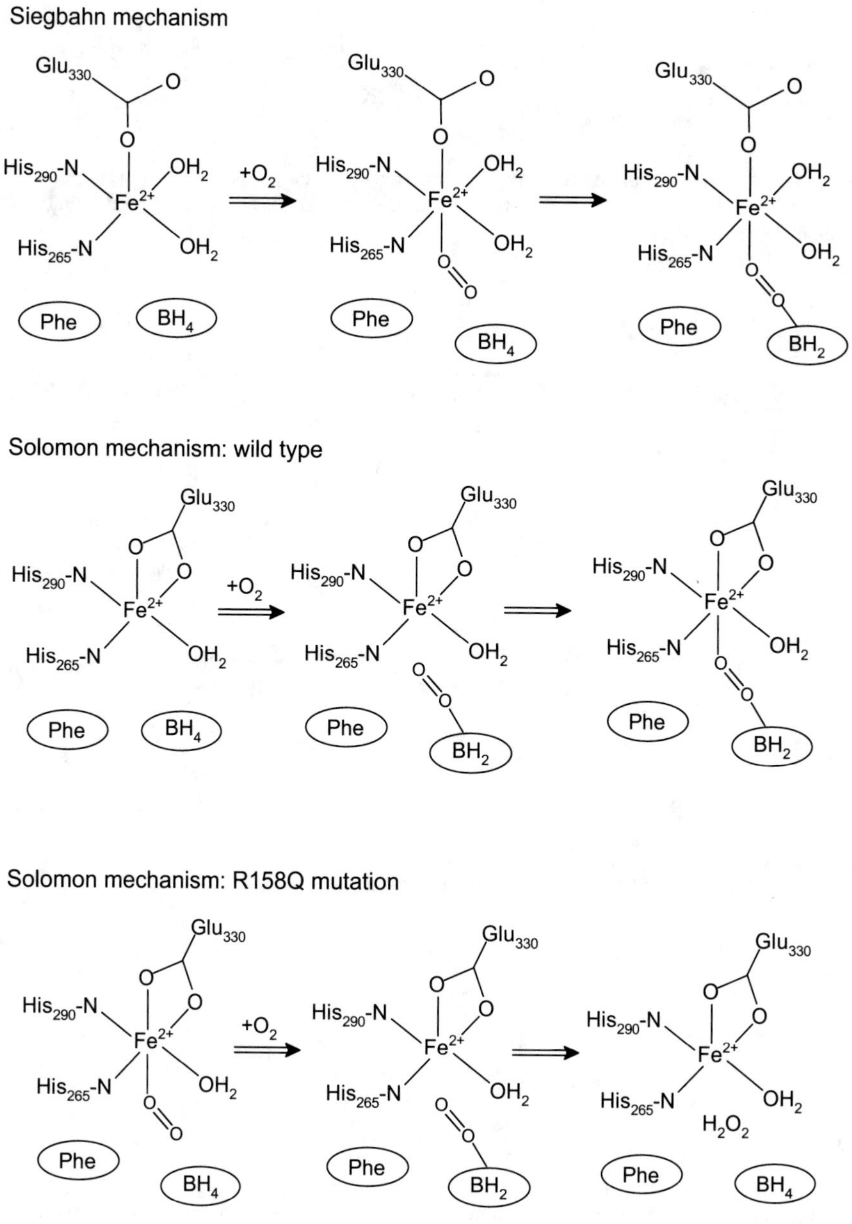

Siegbahn mechanism

Solomon mechanism: wild type

Solomon mechanism: R158Q mutation

Fig. 2.11 Discussed reaction mechanisms of the hydroxylation of phenylalanine (Phe) under the catalytic influence of phenylalanine hydroxylase (PAH). BH4 and BH2 are the redox cofactors (see Fig. 2.12). The active site contains an Fe^{2+} ion coordinated through two histidines and a glutamic acid. Patients with PKU often have the R158Q mutation, which alters the mechanism (Reilley et al. 2019). (Courtesy of the American Chemical Society)

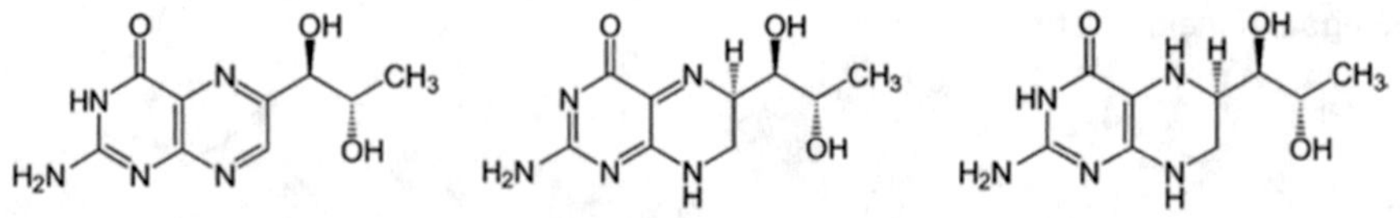

Fig. 2.12 Structures of the basic body biopterin (left) and the redox system dihydrobiopterin (BH2, middle) and tetrahydrobiopterin (BH4, right). 2 double bonds are hydrogenated from biopteridine via BH2 to BH4

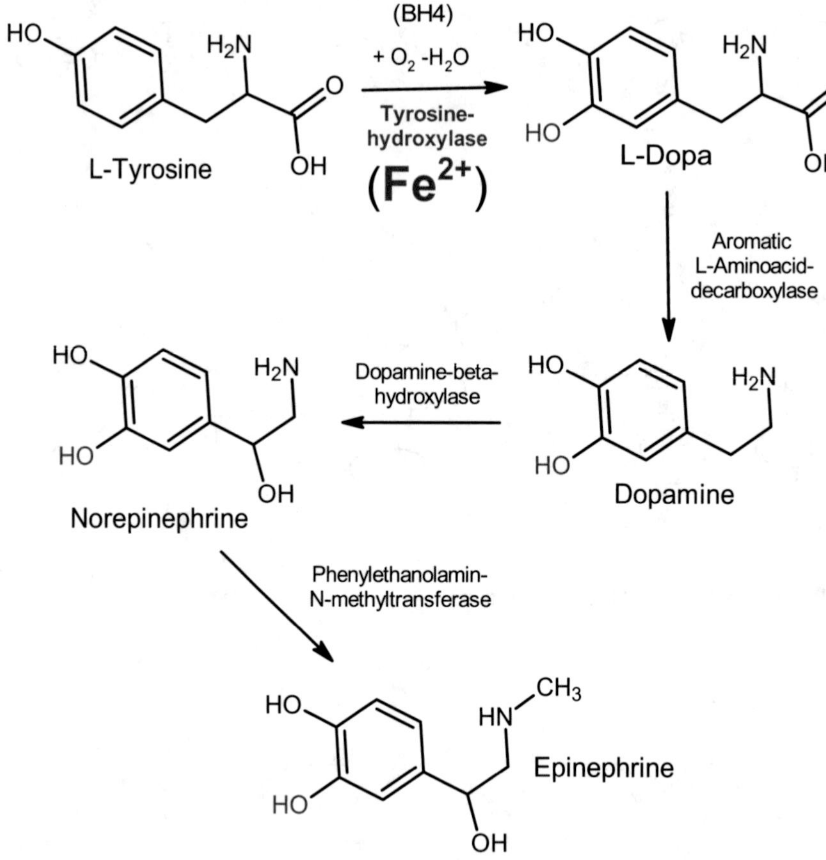

Fig. 2.13 Biosynthetic pathway of catecholamines. Dopamine, norepinephrine and epinephrine are important neurotransmitters and hormones. Iron, as an essential component of tyrosine hydroxylase, plays a crucial role in the ortho-hydroxylation (red) of tyrosine with respect to the first OH group. (Waløen et al. 2017)

Serotonin has a variety of functions in the organism, including in the central nervous system as a transmitter, and has an effect on moods such as serenity and contentment. Thus, depressive moods can partly be associated with a disturbed serotonin metabolism.

In the pineal gland or epiphysis, a part of the diencephalon, melatonin is synthesized and controls, among other things, the day-night rhythm of the human organism. At night, melatonin levels are much higher than during the day. It is used as a
drug against sleep disorders and the release of the hormone is controlled by the eye.
If light falls on the retina, melatonin formation is inhibited.

Serotonin and melatonin are both synthesized in the body from L-tryptophan, an
essential aromatic amino acid for humans with an indole ring system (Fig. 2.14).
L-tryptophan is used as a drug to treat sleep disorders and, above all, makes it easier
to fall asleep. Foods with tryptophan contents >250 mg/100 g include soybeans,
cashew nuts, cheese and lentils.

Hydroxylation of the benzene ring in tryptophan occurs under the catalytic influence of tryptophan hydroxylase, an enzyme that has an Fe^{2+} ion in its active site and
supports a mechanism similar to that of phenylalanine hydroxylase (Fig. 2.11).
After decarboxylation, serotonin is then formed. After acetylation of the amino
group of serotonin and methylation of the OH group, melatonin is finally formed in
the biosynthetic pathway.

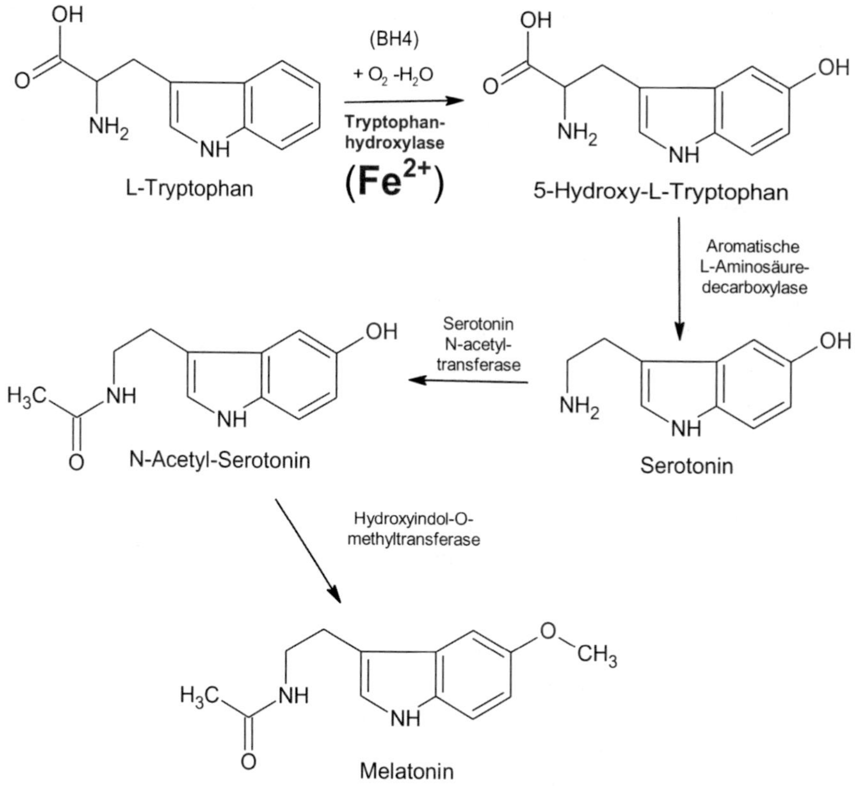

Fig. 2.14 Biosynthetic pathway of the important neurotransmitters and hormones serotonin and
melatonin. Iron plays a crucial role in the hydroxylation (red) of the benzene ring as an essential
component of tryptophan hydroxylase. (Waløen et al. 2017)

The metabolic processes presented here once in depth are intended to show by way of example that iron performs a variety of other central functions in the body in addition to the transport and storage of oxygen. It is involved in important neuro-chemical conversion processes in brain metabolism and in biosynthetic pathways that are responsible for mental characteristics such as motivation, drive and compo-sure, and thus an adequate iron supply is also absolutely essential for human mental health (cf. Figs. 9.3 and 9.4).

The essential function of non-heme iron in neurochemistry is becoming increas-ingly apparent, and the mechanisms of control can be better understood. Thus, the regulation of serotonin synthesis by the iron-containing tryptophan hydroxylase can be influenced by posttranslational modifications such as phosphorylation or cyste-ine oxidation. The discovery of brain-specific isoforms of the enzyme provides fur-ther sophisticated insights into the biochemistry of serotonin formation (Kuhn and Hasegawa 2020).

Iron deficiency during pregnancy can lead to drastic neurological and neuropsy-chological symptoms in the offspring. Sufficient availability of fetal iron influences brain development and function throughout life (Georgieff 2020).

Overall, iron plays an important role in neurodegenerative diseases (Crichton and Ward 2014; Ndayisaba et al. 2019) and the element is also significantly involved in ferroptosis, a specific form of apoptosis (programmed cell death) (Li et al. 2020) (see also Sect. 6.3).

2.7   Iron in the Immune System

Iron plays an important role in the enzyme systems in the innate (non-adaptive), non-specific and acquired (adaptive), specific immune system, both of which are closely intertwined. In leukocytes, such as neutrophils, eosinophils and the T cells, iron is involved in important biochemical mechanisms (Fig. 2.15).

Together with the antibody-producing B cells, T cells are an important compo-nent of the adaptive immune system. As an example, the essential properties of iron for the function of T cells will be presented here.

T cells of the CD8$^+$ type can act directly cytotoxically on cancer cells or infected cells via perforin, which perforates the cell membranes of the target cells and thus makes them permeable to granzymes, which then trigger apoptosis (cell death). The CD4$^+$-T helper cells coordinate the interplay between B cells and the non-adaptive immune system through the release of cytokines (proliferation, differentiation) and chemokines (chemotaxis) (Cronin et al. 2019).

For proliferation and effector activity, T cells require a lot of energy, which is obtained by processes in which heme-iron and iron-sulfur cluster structures (cf. Fig. 2.9) play a very important role. Thus, also one of the first steps in T cell activa-tion and proliferation is the upregulation of the transferrin receptor at the surface of T cells, shortly after T cell receptor (TCR, cf. Fig. 2.16) stimulation (Cronin et al. 2019).

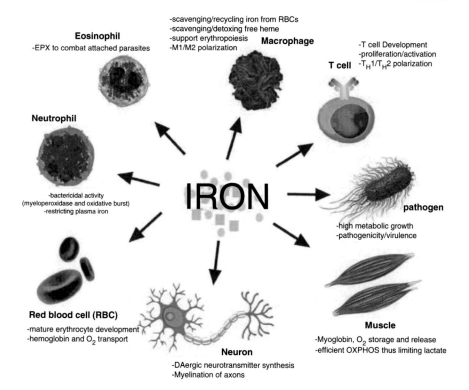

Fig. 2.15 Important functions of iron in the organism. Erythrocytes: oxygen transport. Muscle: oxygen storage and supply, efficient oxidative phosphorylation (OXPHOS), neuron: synthesis of neurotransmitters, myelination (coating with glia cells) of the axon (neuron extension that carries away electrical signals from the cell body). Pathogens deprive the body of iron and are better able to multiply. Therefore, an excess of iron can also promote infection. Neutrophils: about 50% of leukocytes, bacterial defense, part of innate immune defense, control of plasma iron. Eosinophils: about 3% of leukocytes, defense against attached pathogens, EPX is the eosinophil cationic protein. Macrophage: iron recycling from erythrocytes. T cells: Cell development, proliferation and activation, T and B cells are part of the acquired immune response, T_H = T helper cells. (Cronin et al. 2019. Copyright © 2019 Cronin, Woolf, Weis and Penninger. This is an open-access article distributed under the terms of the Creative Commons Attribution License (CC BY))

This illustrates the great importance of iron for the function of the immune system, and therefore iron deficiency and insufficient intake (cf. Figs. 9.3 and 9.4) can lead to a severe impairment of this protective mechanism.

The iron is then required in the mitochondria of the T cells for the production of heme for cytochrome C and for the synthesis of the iron-sulfur clusters in the various complexes of the electron transport chain of oxidative phosphorylation. The import of iron from the cytoplasm into the mitochondria occurs via mitoferrin and can be stored in these organelles as ferritin (Fig. 2.16).

BH4 enhances the iron-dependent mechanisms in T cell activation and the mode of action of the electron transport chain (ETC). The increased metabolic demand

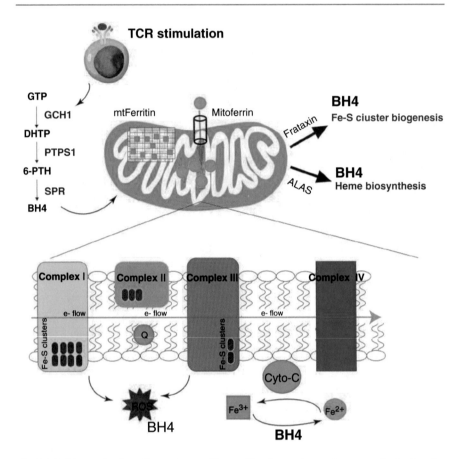

Fig. 2.16 The production of heme-iron and iron-sulfur clusters is very important for the function of T cells in the immune system after their activation at the T cell receptor (TCR). BH4 decisively improves iron-dependent biochemical processes in this process. GTP: guanosine triphosphate, DHTP: dihydroneopterin triphosphate, 6-PTH: 6-pyruvoyltetrahydropterin, BH4: tetrahydrobiopterin (cf. Fig. 2.12). GCH1, PTPS1, SPR: enzymes involved. Frataxin: small protein, plays a crucial role in mitochondrial iron transport. ALAS: enzyme, catalyzes rate-determining step in heme biosynthesis. Mitoferrin: membrane transport protein for Fe^{2+} from cytoplasm to mitochondrion, mt-ferritin: mitochondrial ferritin. Complexes 1–4: components of the respiratory chain. ROS: reactive oxygen species. Cyto-C: cytochrome C. Q: coenzyme Q. (Cronin et al. 2019. Copyright © 2019 Cronin, Woolff, Weis and Penninger. This is an open-access article distributed under the terms of the Creative Commons Attribution License (CC BY))

requires additional regulators for efficient ETC function to limit the generation of the dangerous reactive oxygen species (ROS). To achieve this, expression of the enzyme GCH1 is induced in T cells, which produces the precursor of BH4 upon stimulation. BH4 can not only cause the destruction of ROS, but also directly reduces Fe^{3+} to Fe^{2+}, thus affecting cytochrome C (cyto-C) activity in the ETC (Fig. 2.16) (Cronin et al. 2019).

The example of T cells shows how important an adequate iron supply is for the immune system, and this could be extended to its other components.

T cells of course also play a major role in the body's defense against viral infections, which will be briefly discussed here due to the worldwide relevance of the Corona pandemic starting in the early 2020s.

In late 2019, a coronavirus was identified that has been named SARS-CoV-2 (Severe Acute Respiratory Syndrome-Coronavirus-2) due to the sometimes severe respiratory complications that occur (Wu et al. 2020). Coronaviruses have spike-like glycoproteins on their surface, which are therefore also called S-proteins (spike) and which, when viewed under the electron microscope, appear like a corona around the virus particle.

The virus binds via the S-proteins with high affinity to the angiotensin-converting enzyme 2 (ACE2), which is the target receptor of the host cell, and thus prepares the internalization of the virus particle into the cell. The entry receptor ACE2 is expressed, among others, by epithelial cells of the respiratory tract, but also by enterocytes or gliacells. In the case of SARS-CoV-2 infection, the $CD8^+$ cells discussed also contribute to the immunological defense against the disease and can establish protective immunity by recognizing virus-infected cells by their surface signature and killing them (Ueffing et al. 2020).

This current example shows very clearly how important a sufficient supply of iron to the organism is for the immune system to function optimally. The energy supply for the proliferation and effector activity of the cytotoxic $CD8^+$ cells (killer cells of the adaptive system) can only be ensured in the presence of this important trace element. However, the other components of the immune system also depend on the presence of iron ions, such as the natural killer cells that are part of the innate immune system. To ensure optimal protection, the immune system needs good iron homeostasis. Both a deficiency and an excess of iron can significantly reduce the immune response.

Next to iron, zinc is the most important trace element for the function of the immune system. In contrast to iron, there is no zinc store in the body, and therefore a zinc deficiency can develop very quickly if the intake is insufficient. A large part of the world's population is at high risk of zinc deficiency, which also affects older people in particular and is difficult to determine clinically and chemically because the corresponding storage proteins are missing. Zinc is a component of 300 enzymes and, like iron, is important for the proliferation of the cells of the immune system, which is the most proliferating organ in the body. Furthermore, even in the case of latent deficiency, there are direct effects of zinc on the immune system. For example, the functions of T-, B-cells and natural killer cells (NK, non-adaptive system) are negatively affected, and an overproduction of proinflammatory cytokines takes place. The fact that the T-cell count drops with zinc deficiency because the thymus atrophies is one of the oldest observations in this context (Rink et al. 2018).

The reference values of the D-A-CH societies for zinc have recently been revised and are now given in relation to phytate intake (Haase et al. 2020). Phytic acid (cf. Fig. 7.9) can bind zinc, and thus absorption is impeded. With an average phytate content of foods, the reference values for zinc are 8 mg (w) and 14 mg (m) per day in the age group 19 years and older. Foods naturally rich in zinc include meat, fish, cheese, milk, eggs, nuts, oatmeal, and legumes, although it is always important to consider the phytate content of the food. Oysters are known for their high zinc content.

Since 2012, the D-A-CH societies have regularly revised the reference values for several nutrients according to the latest aspects of science and published them: vitamin D (2012), calcium (2013), folate (2014), selenium (2015), energy intake (2015), vitamin C (2015), vitamin B_1 and B_2 (2016), potassium (2017), sodium and chloride (2017), vitamin B_{12} and protein intake (2019) (Haase et al. 2020).

References

Alenkina I, Kumar A, Berkovsky A, Oshtrakh M (2018) Comparative analysis of the heme iron electronic structure and stereochemistry in tetrameric rabbit hemoglobin and monomeric soybean leghemoglobin alpha using Mössbauer spectroscopy with a high velocity resolution. Spectrochim Acta Part A Mol Biomol Spectrosc 191:547–557

Binnewies M, Finze M, Jäckel M, Schmidt P, Willner H, Rayner-Canham G (2016) Allgemeine und Anorganische Chemie, 3rd edn. Springer Spektrum, Berlin

Crichton R (2016) Iron metabolism: from molecular mechanisms to clinical consequences, 4th edn. Wiley, New York

Crichton R, Ward R (2014) Metal-based neurodegeneration. Wiley, Chichester

Cronin SJF, Woolf CJ, Weiss G, Penninger JM (2019) The role of iron regulation in immunometabolism and immune-related disease. Front Mol Biosci 6:1–19. https://doi.org/10.3389/fmolb.2019.00116

Georgieff MK (2020) Iron deficiency in pregnancy. Am J Obstet Gynecol 223:516–524. https://doi.org/10.1016/j.ajog.2020.03.006

Haase H, Ellinger S, Linseisen J, Neuhäuser-Berthold M, Richter M, on behalf of the German Nutrition Society (DGE) (2020) Revised D-A-CH-reference values for the intake of zinc. J Trace Elem Med Biol 61:126536

Herres-Pawlis S, Klüvers P (2017) Bioanorganische Chemie – Metalloproteine Methoden und Konzepte. Wiley-VCH, Weinheim

Huber C, Wächtershäuser G (1997) Activated acetic acid by carbon fixation on (Fe, Ni) S under primordial conditions. Science 276:245–247

Kaim W, Schwederski B, Klein A (2013) Bioinorganic chemistry – inorganic elements in the chemistry of life, 2nd edn. Wiley, Oxford

Kubo H (1939) About hemoprotein from the root nodules of legumes. Acta Phytochim (Tokyo) 11:195–200

Kuhn MD, Hasegawa H (2020) Chapter 12 – tryptophan hydroxylase and serotonin synthesis regulation. Handb Behavior Neurosci 31:239–256

Lee SC, Lo W, Holm RH (2014) Developments in the biomimetic chemistry of cuban-type and higher nuclearity iron-sulfur clusters. Chem Rev 114:3579–3600

Li J, Cao F, Yin HL, Huang ZJ, Lin ZT, Mao N, Sun B, Wang G (2020) Ferroptosis: past, present and future. Cell Death Dis 11:88. https://doi.org/10.1038/s41419-020-2298-2

Ndayisaba A, Kaindlstorfer C, Wenning GK (2019) Iron in neurodegeneration – cause or consequence? Front Neurosci 13:180. https://doi.org/10.3389/fnins.2019.00180

Reilley D, Popov K, Dokholyan N, Alexandrova A (2019) Uncovered dynamic coupling resolves the ambiguous mechanism of phenylalanine hydroxylase oxygen binding. J Phys Chem B 123:4534–4539

Rink L, Kruse A, Haase H (2018) Immunologie für Einsteiger, 2nd edn. Springer Spektrum, Berlin

Stryer L, Berg JM, Tymoczko JL, Gatto GJ Jr (2018) Stryer Biochemie, 8th edn. Springer Spektrum, Berlin

Ueffing M, Bayyoud T, Schindler M, Ziemssen F (2020) Grundlagen der Replikation und der Immunologie von SARS-CoV-2. Ophthalmologe 117:609–614

Waløen K, Kleppe R, Martinez HJ (2017) Tyrosine and tryptophan hydroxylases as therapeutic targets in human disease. Expert Opin Ther Targets 21:167–180. https://doi.org/10.1080/1472822 2.2017.1272581

Wu F, Zhao S, Yu B, Chen Y, Wang W, Song Z, Hu Y, Tao Z, Tian J, Pei Y, Yuan M, Zhang Y, Dai F, Liu Y, Wang Q, Zheng J, Xu L, Holmes E, Zhang Y (2020) A new coronavirus associated with human respiratory disease in China. Nature 579:265–269

Systemic Iron Homeostasis

3

3.1 Iron Absorption and Recycling

The physiology of iron metabolism in the body is extremely complex, and many processes remain unresolved even today. From a broad perspective, iron metabolism is dominated by two main regulatory systems: In the first, the hormone hepcidin and the iron exporter ferroportin play the main role, and in the second mechanism, iron-regulating proteins are involved that bind to iron-responsive elements of specific messenger RNAs, whose translation in ribosomes then controls, for example, the biosynthesis of ferritin or the transferrin receptor (Henze et al. 2010).

Since the main objective of this book concerns issues from nutrition and food science, the focus is particularly on the relevant processes in the organism. An overview of systemic iron homeostasis is shown in Fig. 3.1 (Muckenthaler et al. 2017).

At the centre of the summary diagram is erythropoiesis, surrounded by the four other areas of enterocyte, macrophage, kidney and liver cells. Iron utilization by the erythroid bone marrow and its recycling by RES macrophages are responsible for the quantitatively most important iron fluxes in the body (Fig. 3.1).

Iron can be absorbed as Fe^{2+} via the divalent metal transporter (DMT-1) and as heme iron directly into the intestinal cells. A new, direct uptake pathway for ferritin iron is independent of the two pathways discussed here (Theil et al. 2012), is not yet shown in Fig. 3.1 and is discussed separately in Sect. 7.3.

Prior to uptake into the intestinal cells via DMT-1, non-heme iron from the diet, if trivalent, is reduced by duodenal cytochrome b (Dcytb) to Fe^{2+} on the apical side. This process will be described in molecular resolution in Sect. 3.2.

Absorption of the heme iron occurs separately with the aid of the heme carrier protein, although the molecular mechanism still needs to be better elucidated. Within the cell, the organic porphyrin system that encloses the divalent iron is oxidatively degraded by heme oxigenase (HMOX1) (Fig. 3.1).

The Fe^{2+} from the two absorption pathways can now be stored as ferritin in the enterocyte after prior oxidation or be discharged via ferroportin (FPN) into the

K. Günther, *Diet for Iron Deficiency*, https://doi.org/10.1007/978-3-662-65608-2_3

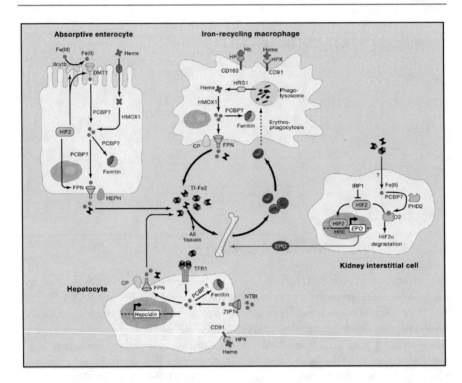

Fig. 3.1 Systemic iron homeostasis: Dcytb (duodenal cytochrome b), DMT1 (divalent metal transporter-1), Heme (heme with heme carrier protein below), PCBP (chaperone), HMOX1 (heme oxigenase 1), Hif2 (hypoxia-induced factor 2), FPN (ferroportin), HEPH (hephaestin), Tf (transferrin), HRG1 (heme transporter), CP (ceruloplasmin ferroxidase), PHD2 (prolyl hydroxylase 2), IRP1 (iron-responsive element protein1), HRE (hormone-responsive element), TFR1 (transferrin receptor 1), ZIP 14 (port for non-Tf bound Fe), HPX (hemopexin), EPO (erythropoietin) (Muckenthaler et al. 2017). (Courtesy of Elsevier)

bloodstream on the basolateral side, where it is bound to transferrin after oxidation to Fe^{3+} with the cooperation of hephaestin. The functional involvement of chaperones (PCBP) is discussed in some processes within the cell. Chaperones are proteins that assist other complex proteins in the correct formation of the three-dimensional structure. Hypoxia-induced factor (HIF2) has a stimulating effect on the translocation of iron through the intestinal cell (Fig. 3.1).

Hypoxia is an undersupply of oxygen to cells. With the help of hypoxia-induced factor (HIF), cells can adapt to a changing oxygen supply; the 2019 Nobel Prize in Medicine was awarded for discoveries in this exciting field (Fig. 3.2). HIF regulates erythropoiesis, among other metabolic pathways. So-called HIF prolyl hydroxylase inhibitors are in development as new and hopeful drugs for renal anemia.

Diferri transferin (Tf-Fe2) supplies iron to cells via the blood, most of which is used to form hemoglobin for red blood cells. Senescent erythrocytes are degraded by specialized macrophages, then the heme iron is directed into the phagolysosome and, after transport by HRG-1 into the cytoplasm, is degraded by the heme oxigenase HMOX1. The released iron is either stored or exported back into the

Fig. 3.2 Gregg Semenza
(b. 1956), professor at
Johns Hopkins University,
USA, received the Nobel
Prize in Medicine in 2019,
together with William
Kaelin and Peter Ratcliffe,
for the discovery of the
molecular mechanisms of
oxygen uptake by cells. (©
Nobel Media. Photo:
A. Mahmoud)

circulation via ferroportin (FPN) and the ferroxidase ceruloplasmin (CP, oxidation to ferric iron) (Fig. 3.1).

When iron is not used, it can be stored in hepatocytes, which can take up diferritransferrin Tf-Fe2 or non-transferrin-bound iron via TFR1 and ZIP14. Hepatocytes store iron as ferritin and they also express FPN and CP for iron export. Heme and hemoglobin (HB) present in the bloodstream interact with hemopexin (HPX) and haptoglobin (HP), respectively, and are secreted via the CD163 and CD91 receptors.

Peritubular fibroblasts of the kidney can detect iron and oxygen deficiency and then release erythropoietin (EPO) to enhance erythropoiesis. When bound to iron and oxygen, prolyl hydroxylase 2 (PHD2) triggers the degradation of HIF2a. Iron loading of PHD2 can be facilitated by poly(rC)-binding proteins (PCBP). Low iron and/or oxygen concentrations inactivate PHD2, leading to HIF2a accumulation and stimulation of EPO transcription. In addition to PHD2, HIF2a is also translationally inhibited by IRP1.

3.2   Molecular Mechanism of Iron Reduction

The biochemical mechanism for the reduction of iron (III) to iron (II) and its absorption is now very well known. As an example of the very deep understanding of the chemistry of iron in bodily functions that is already possible today, the reduction of trivalent food iron by means of duodenal cytochrome b (Dcytb) will be presented in molecular resolution in the following (Fig. 3.3).

Dietary iron absorption is regulated in part by duodenal cytochrome b, an integral membrane protein that catalyzes the reduction of non-heme Fe^{3+} by electron transfer across the membrane from ascorbate.

This step is essential to allow iron uptake by the divalent metal transporter. Figure 3.3 shows the crystallographic structure of human Dcytb and its complexes with ascorbate and Zn^{2+}. Each monomer of the homodimeric protein has cytoplasmic and apical heme groups, and cytoplasmic and apical ascorbate binding sites localized adjacent to each heme. Zn^{2+} is coordinated by two hydroxyl groups of the apical ascorbate and by a histidine residue. Biochemical studies show that Fe^{3+} competes with Zn^{2+} for the binding site at the histidine residue (Figs. 3.3 and 3.4).

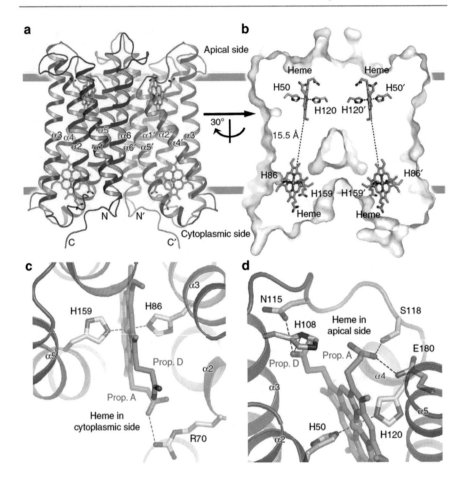

Fig. 3.3 Overall structure of the human Dcytb homodimer. (**a**) Each monomer consists of six α-helices (α1-α6). Both N- and C-termini are located on the cytoplasmic side. Each monomer contains two heme b molecules coordinated by four highly conserved His residues in a six-coordinated low-spin form. (**b**) The distance between two heme molecules is 15.5 Å. (**c**) The environment of heme bound to the cytoplasmic side of Dcytb is shown here. (**d**) The environment of heme bound to the apical side of Dcytb is shown here (Ganasen et al. 2018). (Copyright © 2018 Ganasen, Togashi, Takeda, Asakura, Tosha, Yamashita, Hirata, Nariai, Urano, Yuan, Hamza, Mauk, Shiro, Sugimoto, Sawai. This is an open-access article distributed under the terms of the Creative Commons Attribution 4.0 International License (CC BY): http://creativecommons.org/licenses/by/4.0/)

Based on the structural chemical analysis of duodenal cytochrome b, three possible routes for the transport of electrons supplied by ascorbic acid as a reducing agent (Fig. 3.5) can even be proposed from the cytoplasmic side through the membrane to the apical side (Fig. 3.6). A simplified representation of this important biochemical process in enterocytes for the reduction of iron is shown in Fig. 3.7.

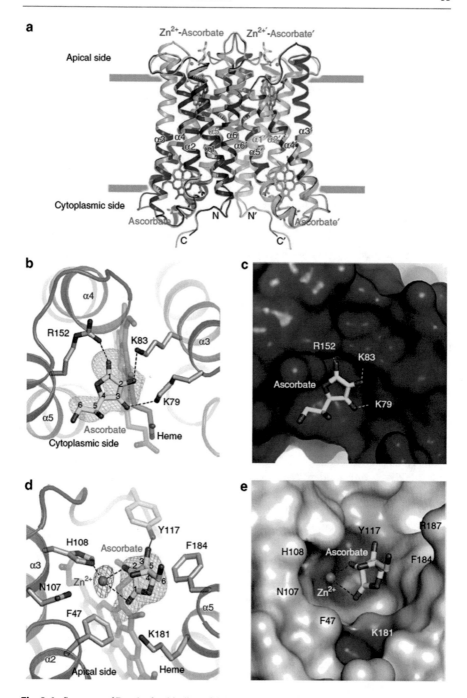

Fig. 3.4 Structure of Dcytb after binding of the substrates ascorbate and zinc. (**a**) The six α-helices of the substrate-free structure (green) were superimposed on those of the substrate-bound structure (blue). (**b**) Ascorbate is bound on the cytoplasmic surface of Dcytb by interaction with three posi-

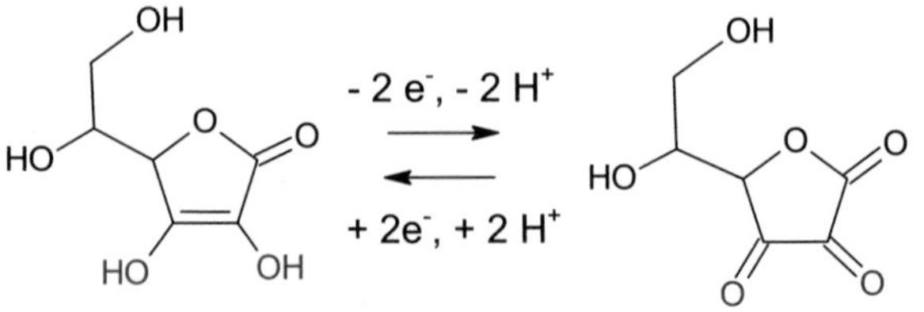

Fig. 3.5 Chemical structure of L-ascorbic acid (left, reducing agent) and dehydroascorbic acid. Two electrons are transferred per molecule

$$- 2\,e^-,\, - 2\,H^+ \longrightarrow$$
$$\longleftarrow + 2e^-,\, + 2\,H^+$$

Taken together, these results provide a very precise structural basis for the mechanism at the molecular level by which Fe^{3+} uptake is promoted by the reducing agent ascorbate. They may thus provide the scientific basis for the development of improved pharmaceuticals for oral iron replacement.

This example shows very clearly the molecular dimensions into which one has already advanced in the physiology of iron resorption. At the biochemical level, there is already a very precise understanding of the processes, and even possible paths of the electrons in the Ångström space for important processes are outlined. For the other functional structural units known to date in iron resorption, such as the divalent metal transporter, heme carrier protein or ferritin port (cf. Fig. 7.11), this enormous structural resolution will certainly also take place in the future and enormously expand the knowledge and options in the supply of iron to the organism in the sense of a molecular nutritional physiology.

3.3   Regulation by Hepcidin

The peptide hormone hepcidin is now regarded as the central regulatory element in iron metabolism (Fig. 3.8). It was discovered during investigations in the context of antimicrobial action (Park et al. 2001), but the connection with iron homeostasis very quickly became evident (Nicolas et al. 2001). The important role of hepcidin in many different processes in the organism continue to evoke much attention recently (Muckenthaler et al. 2017; Camaschella et al. 2020). The diverse signalling pathways regulating hepcidin are shown in Fig. 3.9.

tively charged residues. (**c**) The structure of the cavity for ascorbate binding on the cytoplasmic surface is shown. (**d**) The binding of Zn^{2+} to H108 and two hydroxyl groups of ascorbate on the apical surface of Dcytb is shown. (**e**) The Zn^{2+} – ascorbate-binding cavity on the apical surface is shown here (Ganasen et al. 2018). (Copyright © 2018 Ganasen, Togashi, Takeda, Asakura, Tosha, Yamashita, Hirata, Nariai, Urano, Yuan, Hamza, Mauk, Shiro, Sugimoto, Sawai. This is an open-access article distributed under the terms of the Creative Commons Attribution 4.0 International License (CC BY): http://creativecommons.org/licenses/by/4.0/)

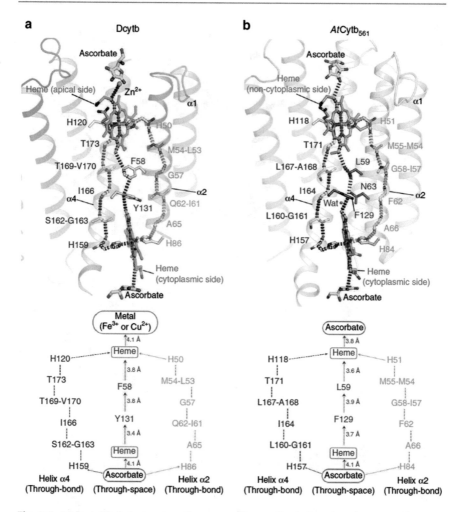

Fig. 3.6 (a) Postulated electron transfer routes of human Dcytb. Based on the present structure, three possible routes can be proposed: First, electron transfer through space (red arrows) mediated by two aromatic residues (Y131 and F58) aligned 3.8 Å apart. Second, electron transfer through the bond (orange dashed line) mediated by a series of amino acid residues along α-helix-2. Third, electron transfer (blue dashed line) mediated by a series of amino acid residues along α-helix-4. (b) Comparison with Cytb from Arabidopsis thaliana (AtCytb$_{561}$) and possible electron transfer routes (Ganasen et al. 2018). (Copyright © 2018 Ganasen, Togashi, Takeda, Asakura, Tosha, Yamashita, Hirata, Nariai, Urano, Yuan, Hamza, Mauk, Shiro, Sugimoto, Sawai. This is an open-access article distributed under the terms of the Creative Commons Attribution 4.0 International License (CC BY): http://creativecommons.org/licenses/by/4.0/)

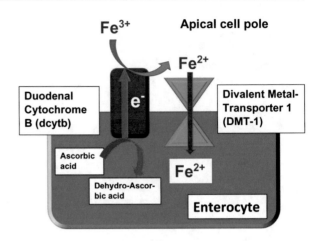

Fig. 3.7 Simplified
representation of the
reduction of ferric iron to
ferrous iron by cytosolic
ascorbic acid and
subsequent uptake of the
ferrous iron in the
enterocyte. The electrons
are transported via a
transmembrane process
through Dcytb

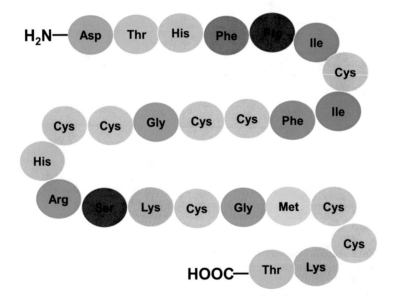

Fig. 3.8 Primary structure of hepcidin-25, the central regulator of iron biochemistry in the human organism

In the presence of filled iron stores, BMP6 (bone morphogenetic protein 6) is formed. Together with its co-receptor hemojuvelin (HJV), type 1 (Alk2/3) and type 2 (BMPR2, ACVR2A) BMP serine-threonine kinase receptors are activated. This leads to phosphorylation of receptor-activated SMAD (R-SMAD) proteins and formation of active transcription complexes with SMAD4. The interaction between HJV and BMP can be facilitated by neogenin (Fig. 3.9).

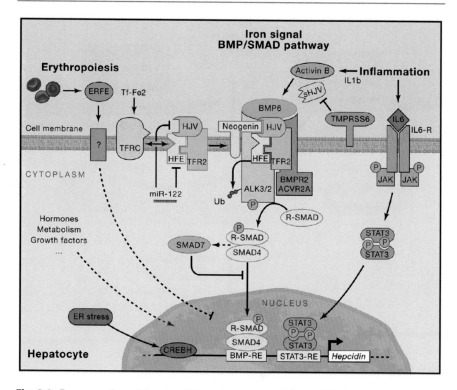

Fig. 3.9 Representation of the signalling pathways in which hepcidin is regulated. Hepcidin expression is regulated by iron signaling (in the middle), erythropoiesis (left), and inflammatory processes (right) (Muckenthaler et al. 2017). (Courtesy of Elsevier)

High concentrations of Tf-Fe2 displace HFE from the complex with TFR1. HFE then forms an association with TFR2 and HJV, which then promotes BMP/SMAD signaling to hepcidin. HFE additionally interacts with ALK3 and prevents its ubiquitination (Ub) and degradation. The BMP/SMAD pathway to hepcidin is suppressed by serine protease matriptase 2 (TMPRSS6), which cleaves HJV to produce a soluble form of HJV (sHJV). This pathway is inhibited by SMAD7 and by the microRNA miR-122, which decreases HJV and HFE expression (Fig. 3.9).

In inflammation, JAK tyrosine kinase is activated by the interaction of interleukin 6 (IL-6) with its receptor (IL-6R). The kinase triggers the formation of the STAT3 complex (signal transducer and transcription activator 3), which binds to the hepcidin promoter in the nucleus. Stimulation of hepcidin formation by activin B is dependent on the BMP/SMAD pathway. The STAT3 and BMP-responsive elements (REs) in the hepcidin promoter are shown explicitly (Fig. 3.9).

High erythropoietic activity increases plasma levels of ERFE, which then suppresses hepcidin. The biochemical pathways involved are not yet known. Other

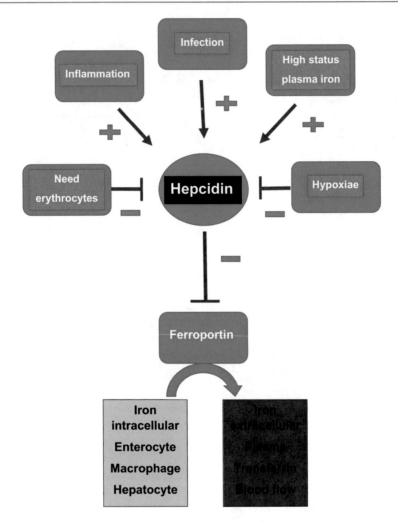

Fig. 3.10 Simplified representation of the regulation of the synthesis of hepcidin and the effect on ferroportin, which exports iron to the extracellular space

factors may modulate hepcidin expression. These include protein misfolding (ER stress), which activates hepcidin transcription via CREBH, and various hormones or growth factors (Fig. 3.9).

The most important influences on hepcidin levels are clearly summarized in Fig. 3.10. Inflammation, infection and a high status of plasma iron increase the hepcidin level, while a requirement for iron in erythropoiesis and a lack of oxygen reduce the concentration of the peptide.

A high concentration of hepcidin has an inhibitory effect on the function of ferroportin, which is responsible for the transport of intracellular iron into the

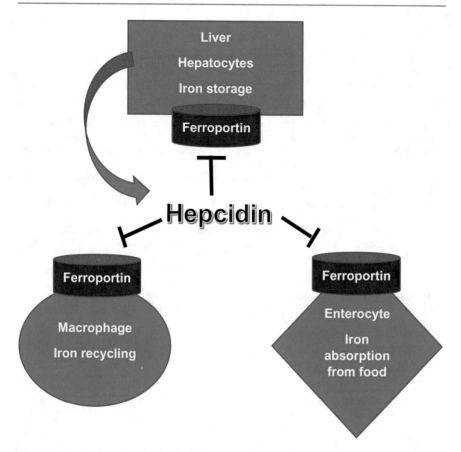

Fig. 3.11 Synthesis and main actions of hepcidin. After formation in the liver, it has an inhibitory effect on the function of ferroportin in the enterocytes, macrophages and hepatocytes

bloodstream and thus plays an extremely important role in the iron cycle of the organism (Figs. 3.10 and 3.11). When hepcidin binds to ferroportin, it is transported into the cell interior and degraded there.

The serum level of iron in humans is thus tightly controlled by the action of the hormone hepcidin on the iron transporter ferroportin. Abnormal ferroportin activity can also lead to diseases such as iron overload (hemochromatosis). Hepcidin is also becoming increasingly important in the diagnosis of iron deficiency states and distribution disorders and is regarded as the central regulator in the entire iron metabolism (cf. Chap. 5).

Due to the great importance of this peptide, its biochemical function and effect on ferroportin in particular is the subject of in-depth investigations at the molecular level. With the chemical function models derived from this, targeted intervention in the hepcidin-ferroportin interaction in pathological disorders of iron homeostasis will become possible in the future (Billesbølle et al. 2020).

References

Billesbølle C, Azumaya C, Kretsch R, Powers A, Gonen S, Schneider S, Arvedson T, Dror R, Cheng Y, Manglik A (2020) Structure of hepcidin-bound ferroportin reveals iron homeostatic mechanisms. Nature 586:807–811

Camaschella C, Nai A, Silvestri L (2020) Iron metabolism and iron disorders revisited in the hepcidin era. Haematologica 105:260–272

Ganasen M, Togashi H, Takeda H, Asakura H, Tosha T, Yamashita K, Hirata K, Nariai Y, Urano T, Yuan X, Hamza I, Mauk G, Shiro Y, Sugimoto H, Sawai H (2018) Structural basis for promotion of duodenal iron absorption by enteric ferric reductase with ascorbate. Nature Commun Biol 1:120. https://doi.org/10.1038/s42003-018-0121-8

Henze M, Muckenthaler M, Galy B, Camaschella C (2010) Two to tango: regulation of mammalian iron metabolism. Cell 142:24–38

Muckenthaler M, Rivella S, Hentze M, Galy B (2017) A red carpet for iron metabolism. Cell 168:344–361

Nicolas G, Bennoun M, Devaux I, Beaumont C, Grandchamp B, Kahn A, Vaulont S (2001) Lack of hepcidin gene expression and severe tissue iron overload in upstream stimulatory factor 2 (USF2) knockout mice. Proc Natl Acad Sci U S A 98:8780–8785

Park C, Valore E, Waring A, Ganz T (2001) Hepcidin, a urinary antimicrobial peptide synthesized in the liver. J Biol Chem 276:7806–7810

Theil E, Chen H, Miranda C, Janser H, Elsenhans B, Nunez M, Pizarro F, Schümann K (2012) Absorption of iron from ferritin is independent of heme iron and ferrous salts in woman and rat intestinal segments. J Nutr 142:478–483

Anemia and Iron Deficiency

4

Already some centuries ago the lack of iron and the associated anaemia was described as pallor or chlorosis. Young women in particular were affected and the correct diagnoses were not always made in the past. Anemia has negative consequences for the entire organism and can have a very strong negative impact on the whole life (Fig. 4.1).

Iron metabolism plays an important role in very many different bodily functions, and iron deficiency can therefore often cause several symptoms and complaints at the same time, which should be carefully observed. In total, one can name ten important signs that can indicate iron deficiency:

- Tiredness, general feeling of weakness, paleness of the skin and mucous membranes, easy exhaustibility, reduced performance and resilience at work.
- Nervousness, concentration disorders with forgetfulness
- Emotional lability and depressed moods
- Morning headache
- Increased susceptibility to infections and sensitivity to cold
- Appetite disorders and constipation
- Brittle nails, hollow nails, brittle hair, hair loss
- In children growth and development disorders
- Recent studies also show a link between iron deficiency and heart disease
- Restless legs syndrome (RLS), urge to move the legs that occurs only when resting or relaxing, usually associated with sensory disturbances or pain.

Fig. 4.1 Patients feel severely impaired in their overall lifestyle. Fatigue, listlessness and even depression can be the consequences of iron deficiency and anaemia. According to the definition of the World Health Organization (WHO), one speaks of anemia when the hemoglobin concentration (Hb value) falls below 12 g/dl in women and 13 g/dl in men. (© GoodIdeas/stock.adobe.com)

Especially in risk groups, an iron deficiency can develop quickly. Therefore, special attention should be paid here to the symptoms mentioned. These are especially pregnant women and women with heavy menstrual bleeding, children, competitive athletes and elderly people with concomitant diseases, especially of the kidneys, patients with chronic infections and diseases of the gastrointestinal tract, heart disease and patients with weakened immune defenses.

For an initial assessment, one should go through the ten symptoms listed above. If several of the points apply, a laboratory examination of the iron values in the blood should be ordered in order to find out whether the complaints can be attributed to an iron deficiency or whether other causes play a role (cf. Chap. 5).

4.1   Global Issues

Figure 4.2 shows the global problem of iron deficiency using the example of anaemia in children (WHO 2015). Anemia is not always due to iron deficiency, but in about 80% of cases. In some developing countries in particular, the proportion of people with anaemia is very high. Reasons for this include a severe lack of food of all kinds and often additionally the poor quality of the food consumed there. This is

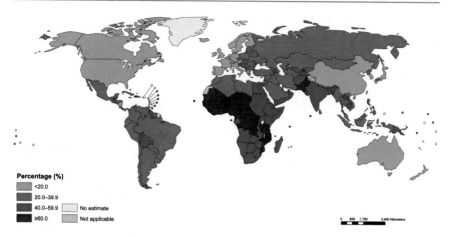

Fig. 4.2 Percentage (prevalence) of children with anemia aged 0.5–5 years worldwide in 2011 (World Health Organization 2015). In addition to hunger, "hidden hunger" is also a global problem. Food with optimized composition and sufficient supply could help here

often referred to as "hidden hunger" and means that indispensable nutrients such as iron, other trace elements and minerals as well as vitamins, essential fatty and amino acids and other important substances are missing from the food.

A very large proportion of the world's population, an estimated 3 billion people, eat food of inferior quality, and this affects inhabitants of all countries of the world. Particularly frequent deficiencies are iron, zinc, vitamin B_9 (folic acid) and vitamin A.

Due to the future higher CO_2 concentration of the atmosphere, the contents of iron and zinc in plant foods will be reduced, and thus a further reduction in the quality of these trace elements will occur. Thus, it supply of important micronutrients is also related to climate change. Several studies have investigated the influence of CO_2 concentration in the atmosphere on the concentration of iron and zinc in the edible portions of the globally important food crops rice, wheat, maize, soybean and pea. For all 5 foods, a decrease in the content of both elements of about 4–10% was found at a CO_2 level of about 550 ppm (ppm: parts per million) in the ambient air. The protein concentration also decreases (Myers et al. 2014).

At present, the CO_2 content of the atmosphere is around 410 ppm, and a further increase is expected, the quantity of which, however, is not easy to forecast. Many experts assume an increase to 550 ppm in the next 40–60 years.

In addition, the implications of these results for the nutritional situation of the world's population were analysed. The results showed that increased CO_2 levels could lead to an additional 175 million people suffering from zinc deficiency and an additional 122 million people suffering from protein deficiency (assuming population and CO_2 projections for 2050). In the case of iron, there are currently 1.4 billion women of childbearing age and children under 5 living in countries with anemia prevalence greater than 20%. These would lose >4% of the iron content in their diet.

The regions most affected in this regard are South and Southeast Asia, Africa, and the Middle East (Smith and Myers 2018).

With this in mind, future solutions within a modern and global bioeconomy will need to be developed using new methods in plant science to ensure food security in terms of value-added ingredients globally and for future generations. Current knowledge on the specifications for a globally ideal food basket has recently been linked to the essential considerations for a healthy and sustainable world diet in the Anthropocene (Willett et al. 2019).

The Anthropocene is the term used to describe the period in Earth's history when humans began to exert a significant influence on all systems on our planet. The year 1950 is cited by most experts as the time of onset. The study describes an optimized reference food basket for the sustainable feeding of the world's population for the years up to and from 2050, when 10 billion people are expected to live on Earth. The optimized food combination is precisely quantified in the publication and consists of vegetables, fruits, whole grains, nuts, unsaturated oils and moderate amounts of fish and poultry. If possible, red and processed meats, added sugars, refined grains, and starchy products should be avoided. This reference food basket, which also focuses on micronutrients such as iron, should apply to all countries and cultures (Willett et al. 2019).

In connection with "hidden hunger," ultra-processed foods, which now account for a large share of the world's food supply and which often lack important micro-nutrients, are also repeatedly discussed (Haddad et al. 2016). In relatively high-income countries such as the USA, Canada and the UK, more than half of total dietary energy is already consumed through ultra-processed foods. In middle-income countries such as Brazil, Mexico or Chile, the corresponding share is between one-fifth and one-third of the kilocalories consumed. The average market growth for these products is 1% per year in high-income countries and up to 10% in middle-income countries. Ultra-processed foods are typically products with high energy density, high levels of sugar, unflavorful fats and table salt, and low levels of fiber, protein, vitamins and minerals, and trace elements. Based on numerous studies, they are considered to be the cause of a wide range of diseases (Monteiro et al. 2019).

4.2   Other Causes

In addition to iron deficiency anaemia (sideropenic anaemia, 80% of all cases), however, there are other deficiency states that cause a disturbance of erythrocyte formation and can thus lead to anaemia. These include vitamin B_{12} deficiency anemia or pernicious anemia. William Parry Murphy, George Minot and George Whipple were awarded the Nobel Prize in Medicine in 1934 for the liver therapy of pernicious anaemia (Figs. 4.3 and 4.4).

Furthermore, there are forms of anaemia that are due to a deficiency of protein, vitamin B_6, vitamin C or folic acid (Fig. 4.5).

Fig. 4.3 Typical sources
of vitamin B_{12} are liver,
meat, cheese or chicken
eggs. Beef liver, for
example, contains 65 µg
vitamin B_{12}/100 g (see also
Sect. 9.1). (© AlionaUrsu/
stock.adobe.com)

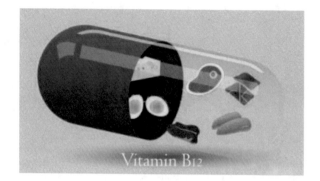

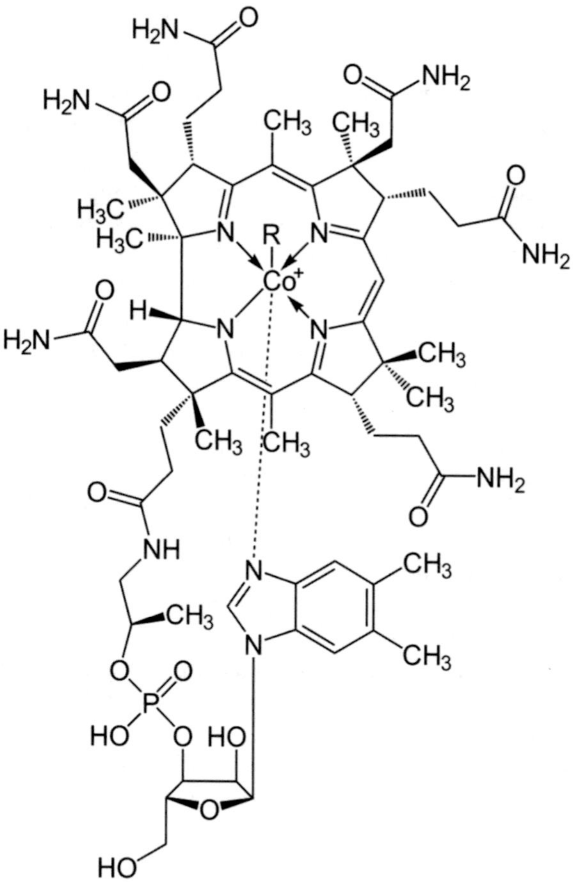

Fig. 4.4 Vitamin B_{12} is a
collective name for cobalt
complexes and is
particularly important for
blood formation, cell
division and the function
of the nervous system. The
bioavailability in the
organism depends on the
residue R bound to the
cobalt. Therapeutically
used preparations are
cyanocobalamin (R = CN)
or hydroxycobalamin
(R = OH). The estimated
value for an adequate
intake for adults is given
by the German Society for
Nutrition as 4 µg/day

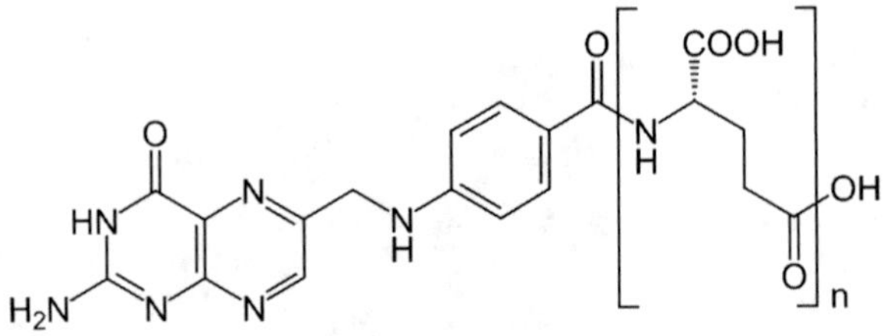

Fig. 4.5 Folic acid is a heat- and light-sensitive vitamin and is composed of a pteridine-p-aminobenzoic acid and an L-glutamate moiety ($n = 1$ folic acid, $n > 1$ polyglutamates of folic acid). Folic acid is the precursor of the coenzyme tetrahydrofolic acid, which is required for the synthesis of purine bases and is thus essential for cell division. For adults, the German Society for Nutrition recommends an intake of 300 µg folate equivalent/day. 1 g folate equivalent = 1 g dietary folate = 0.5 g synthetic folic acid

Renal anaemia occurs when too little erythropoietin (EPO) is produced by the kidneys. EPO is a messenger substance for blood formation and tells the bone marrow to form erythrocytes. In certain kidney diseases, EPO production is limited, too few red blood cells are formed, and anemia is the result.

Furthermore, the following causes can also be responsible for anemia:

Loss of blood from the body due to sudden bleeding, e.g. during surgery, or chronic bleeding due to stomach ulcers or intestinal polyps.

Autoimmune diseases, infections and other diseases can hinder the formation of erythrocytes.

In aplastic anemia, which is very rare, the bone marrow may produce few or no erythrocytes.

Erythrocytes are destroyed or degraded prematurely (haemolytic anaemia), sickle cell anaemia or thalassaemia (so-called Mediterranean anaemia), here a defect in the genetic information causes the formation of sickle-shaped erythrocytes which are destroyed more quickly (cf. Sect. 2.2), poisoning by heavy metals (lead, copper, enzymes are inhibited), malaria, the pathogens multiply in red blood cells and destroy them.

Erythrocytes are not evenly distributed in the body, this can happen, for example, if the spleen is greatly enlarged.

In summary, the literature shows that iron deficiency is the cause of 80% of all cases of anaemia and that a special focus on the body's iron supply and iron biochemistry is therefore entirely justified. However, combination effects can also play a role, e.g. if the diet includes foods of reduced quality (hidden hunger, low micronutrient density, e.g. little iron, B_{12}, folate at the same time).

References

Haddad L, Hawkes C, Waage J, Webb P, Godfray C, Toulmin C (2016) Food systems and diets: facing the challenges of the 21th century. Global Panel on Agriculture and Food Systems for Nutrition. City, University of London Institutional Repository, London

Monteiro C, Cannon G, Levy R, Moubarac JC, Louzada M, Rauber F, Khandpur N, Cediel G, Neri D, Martinez-Steele E, Baraldi L, Jaime P (2019) Ultra-processed foods: what they are and how to identify them. Public Health Nutr 22:936–941

Myers S, Zanobetti A, Kloog I, Huybers P, Leakey A, Bloom A, Carlisle E, Dietterich L, Fitzgerald G, Hasegawa T, Holbrook N, Nelson R, Ottman M, Raboy V, Sakai H, Sartor K, Schwartz J, Seneweera S, Tausz M, Usui Y (2014) Rising CO_2 threatens human nutrition. Nature 510:139–142

Smith M, Myers S (2018) Impact of anthropogenic CO_2 emissions on global human nutrition. Nat Clim Chang 8:834–839

Willett W, Rockström J, Loken B, Springmann M, Lang T, Vermeulen S, Garnett T, Tilman D, DeClerck F, Wood A, Jonell M, Clark M, Gordon L, Fanzo J, Hawkes C, Zurayk R, Rivera J, De Vries W, Sibanda L, Afshin A, Chaudhary A, Herrero M, Agustina R, Branca F, Lartey A, Fan S, Crona B, Fox E, Bignet V, Troell M, Lindahl T, Singh S, Cornell S, Reddy S, Narain S, Nishtar S, Murray C (2019) Food in the anthropocene: the EAT-lancet commission on healthy diets from sustainable food systems. Lancet 393:447–492

World Health Organization (2015) The global anaemia prevalence in 2011. WHO, Genf

Diagnostics of Iron Deficiency States

5

5.1 Basic Diagnostics

To determine an iron deficiency, a blood sample is taken in which the various parameters are then examined. From the interpretation of these results, a possible iron deficiency can be deduced.

The main features of iron diagnostics consist of the evaluation of four values, from which, as a rule, a statement can be made quite reliably.

In the case of complicated conditions such as concomitant diseases, further values should be determined in the blood to provide a more accurate view of the conditions. These parameters and recent developments are discussed in Sect. 5.2.

First and foremost, it is important to note: In the diagnosis of iron deficiency, the sole determination of iron in the blood serum, i.e. in the liquid portion of the blood sample, is still occasionally used in practice today, but this only allows an uncertain statement on the iron status of a patient. This is because the determination of total iron in serum is highly dependent on the time of day and other influencing factors and is therefore not suitable for diagnostic purposes. This is called circadian rhythm. The value tends to be low at night and very high in the afternoon. The total concentration is also high after a meal rich in iron. On the basis of a low total iron value in the serum, iron substitution (i.e. taking medicines with iron) must therefore not be started, even if other findings of the blood count "fit" with this (Schrezenmeier 2011).

Crucial and important for determining the iron supply, on the other hand, are the determination of at least four values: hemoglobin (Hb value), transferrin saturation (iron transporter), ferritin (iron store) and C-reactive protein (CRP, inflammation) (Fig. 5.1). Some of these values are often not part of a normal iron deficiency examination and must be requested separately by the physician.

© The Author(s), under exclusive license to Springer-Verlag GmbH, DE, part of
Springer Nature 2023
K. Günther, *Diet for Iron Deficiency*,
https://doi.org/10.1007/978-3-662-65608-2_5

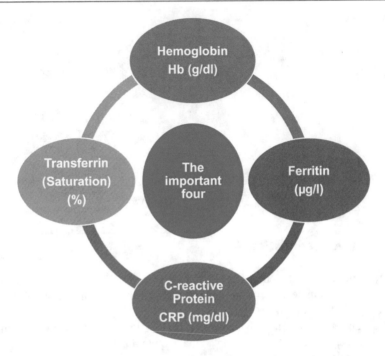

Fig. 5.1 The four important laboratory parameters for determining iron status. Determination of the Hb value alone is not sufficient

In most doctors' practices, the iron status of a patient is initially assessed only by the hemoglobin value (Hb value). However, this is only part of the truth, because the Hb value says little about the filling of the iron stores (ferritin) and the percentage loading of the transferrin with iron.

Thus, a normal Hb value can feign an adequate supply, while, for example, there is only a low transferrin saturation (iron deficiency without anemia). As a result, the many different iron enzymes can no longer be sufficiently supplied with the trace element and therefore no longer function properly, and illnesses can develop, including depression (cf. Sects. 2.5 and 2.6).

With the Hb value, the amount of hemoglobin is usually displayed in the unit (g/dl). The Hb value only starts to decrease when the iron stores (ferritin) have already been emptied. An iron deficiency was therefore already present beforehand. This has now developed into iron deficiency anemia. The serum ferritin value is given in the unit µg/l and indicates how full the iron stores are. However, ferritin is an acute-phase protein, and this means that when there is inflammation in the body, ferritin levels are elevated, thus masking any storage iron deficiency that may be present.

Acute-phase proteins are proteins that appear in increased numbers in the blood in the event of tissue damage or infection, among other things as part of the non-specific immune reaction (acute-phase reaction). Their synthesis is induced by various interleukins. Inflammation can be determined by the C-reactive protein (CRP), which is also an acute-phase protein and is the classic inflammation parameter. CRP got its name because it can bind to the C-polysaccharide of the cell wall of Streptococcus pneumonia (causes pneumonia). As a marker for phagocytosis (ingestion and destruction of foreign substances), CRP also attaches to other bacteria, fungi and parasites. The phagocytes then eliminate the complex with CRP and the specific antigen.

If inflammation is present with CRP values >5 mg/l, the transferrin saturation can also be considered, which is given as the loading of the protein with iron in percent and is little influenced by inflammation. In patients with chronic disease, this plays a major role and must be closely monitored. In severe infections, CRP values can be in the range of up to 400 mg/l.

In the normal range, the transferrin saturation is between 20% and 45%. If the glycoprotein is loaded with <20% iron, the supply of the organism is no longer optimal. This can also be the case when the Hb value is still in the reference range. An iron deficiency situation – with its many disadvantages – would not be detected at all by the Hb value alone.

The reference values for hemoglobin and ferritin as a function of age and sex are listed in Table 5.1. The wide range of variation in the values for ferritin is particularly striking. One reason for this is the different determination methods used in routine clinical chemistry and laboratory medicine. Here, too, the iron content is not targeted, but the ferritin proteins are used as analytical targets, which surround the iron core, which is more or less large depending on the loading (cf. Fig. 2.6). In the immunoassays used, the proteins are bound to antibodies of different quality and these complexes are then detected in different ways. Therefore, one and the same

Table 5.1   Reference values of the diagnostic parameters of iron metabolism (values according to Schrezenmeier 2011)

Population	Hemoglobin [g/dl]	Population	Serum ferritin[a] [µg/l]
Newborn (1–4 days)	16.2–21.2	Newborn (2 weeks)	90–628
Newborn (1–2 weeks)	15.5–19.6	Baby (1 month)	144–399
Baby (2–4 weeks)	12.6–17.2	Baby (2 months)	87–430
Baby (5–12 weeks)	10.5–12.6	Baby (4 months)	37–233
Baby (>12 weeks)	11.0–14.4	Baby (6 month)	19–142
Children (6 month – 5 years)	11.0	Baby (9 month)	14–103
Children (5–11 years)	11.5	Baby (12 month)	11–91
Children (12–14 years)	12.0	Children (1–10 years)	15–119
Women (>15 years, not pregnant)	12.0	Women (20–50 years)	23–110
Pregnant women	11.0	Women (65–90 years)	13–651
Women in postpartum	10.0	Men (20–50 years)	35–217
Men (>15 Jahre)	13.0	Men (65–87 years)	4–665

[a]Values can be method dependent

blood sample does not always give the same result when using different assays. Therefore, the reference ranges for each determination method must be considered separately.

Using species-specific isotope dilution mass spectrometry, the amount of iron bound in ferritin can be determined very precisely (Hoppler et al. 2009). This method therefore has great advantages over immunoassays, but requires a costly mass spectrometer and the use of analytical chemistry specialists. For this reason, it has not yet played a role in routine determination in everyday clinical practice. This method is discussed in more detail in Chap. 7.

According to the definition of the World Health Organization (WHO), one speaks of anemia when the Hb value falls below 12 g/dl in women or below 13 g/dl in men.

5.2   Further Diagnostics

If the results of the examinations, even of other types, are not conclusive in their overall picture, the physician must consult additional blood values in order to reliably diagnose an iron deficiency. An important parameter in this context is the mean corpuscular volume of the erythrocytes (MCV) (Fig. 5.2), which is routinely examined as part of the so-called small blood count. Eight parameters are determined in the small blood count: Erythrocytes, leukocytes, platelets, hematocrit (Hkt), hemoglobin (Hb), mean corpuscular hemoglobin (MCH), mean corpuscular hemoglobin concentration (MCHC), mean corpuscular volume (MCV) of the erythrocytes.

The MCV value belongs to the group of three erythrocyte indices which indicate the average volume and the average hemoglobin content and concentration. Depending on the level of the MCV value, a macrocytic, normocytic or microcytic anemia can be distinguished (Figs. 5.2 and 5.3).

In macrocytic anemia the erythrocytes are enlarged and in the case of the microcytic form they are reduced in size compared to the normal size (normocytic). Based on the MCV value, possible causes in various forms of anemia can now be narrowed down (Fig. 5.4). For example, iron deficiency anemia, which accounts for 80% of

Fig. 5.2 Differential diagnosis of anemia according to the mean corpuscular volume (MCV) of the erythrocytes in femtoliters (fl). 1 fl is 10^{-15} l. (Values according to Schrezenmeier 2011)

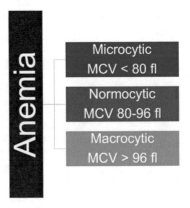

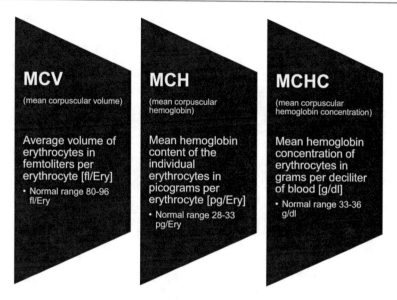

Fig. 5.3 The three important erythrocyte indices MCV, MCH and MCHC with their units and normal ranges. (Values according to Schrezenmeier 2011)

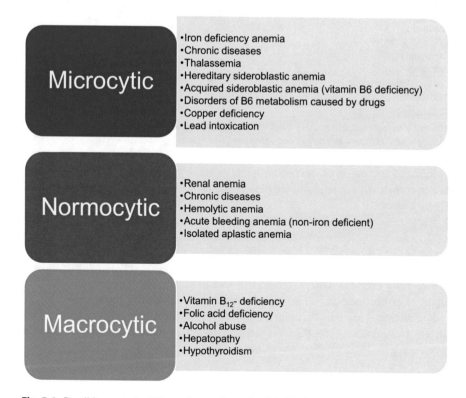

Fig. 5.4 Possible causes in different forms of anemia. (Modified after Schrezenmeier 2011)

Table 5.2 Differential diagnosis of various disorders of iron metabolism with the aid of laboratory parameters

	Hemo-globin (Hb)	Ferri-tin	Trans-ferrin saturation	Soluble trans-ferrin receptor (sTfR)	Hb content retiku-locytes (CHr)	Zinc proto-porphy-rin (ZPP)	Hepci-din levels expec-ted
Iron deficiency anemia	↓	↓	↓	↑	↓	↑	↓
Storage iron deficiency	N	↓	N	N	N	N	N
Iron-deficient erythropoiesis due to iron deficiency	N	↓	↓	↑	↓	↑	↓
Iron-deficient erythropoiesis due to iron distribution disorders in chronic diseases	N/↓	N/↑	↓	N	↓	↑	↑
Iron deficiency and chronic disease	N/↓	N/↓	↓	N/↑	↓	↑	–
Resistance to erythropoiesis-stimulating substances due to functional iron deficiency	↓	N/↓	↓	↑	↓	↑	↑
Anemia with ineffective erythropoiesis	↓	↑	↑	↑	–	–	↓

Modified after Schrezenmeier (2011); red arrow pointing downwards means decrease, green arrow pointing upwards means increase of the corresponding values. N = normal, – variable

all forms of anemia, is associated with microcytic erythrocytes, while renal disease causes normocytic anemia, and the macrocytic form is formed in the case of vitamin B_{12} deficiency.

Further parameters of a differential diagnosis are the soluble transferrin receptor (sTfR), the hemoglobin content of the reticulocytes (CHr) and the zinc protoporphyrin (ZPP) (Table 5.2).

Transferrin receptors are glycoproteins, are located on the cell membrane and are continuously shed from there. The sTfR value in serum is an indicator for the iron supply of erythropoiesis and is not influenced by inflammatory conditions.

Reticulocytes are the precursor cells of erythrocytes. Checking the hemoglobin content of the reticulocytes (CHr) is therefore indicated in the case of a questionable latent iron deficiency and when checking a substitution of iron that has been carried out. Only less than 1% of erythrocytes are replaced daily. Therefore, in the case of latent iron deficiency, only the youngest erythrocytes would be affected. With a red blood cell life span of 120 days, it would therefore take a very long time to detect a latent iron deficiency by pathological values of MCV, MCM or MCHC (cf. Fig. 5.3), which would cause a manifest deficiency.

Zinc protoporphyrin (ZPP) is formed when the divalent Fe^{2+} is replaced by the divalent Zn^{2+} in protoporphyrin IX. The enzyme ferrochelatase incorporates Fe^{2+} into protoporphyrin IX in the final phase of heme synthesis. In very small amounts, however, bivalent zinc is also bound at the same site instead of iron in healthy individuals. In the case of iron deficiency, the proportion of ZPP naturally increases, but also, for example, in the case of intoxication with lead, which inhibits ferrochelatase. The ZPP determination therefore detects iron-deficient erythropoiesis.

The determination of the concentration of the peptide hepcidin (cf. Sect. 3.3) in serum, which is central to iron regulation, makes it possible to distinguish absolute iron deficiency (reduced hepcidin value) from functional iron deficiency in chronic diseases (increased hepcidin value). So far, however, there is still work to be done in standardizing the measurement of this relatively new parameter. The focus is also currently only on the determination of the isoform hepcidin-25, and various immunochemical methods are being established. However, the gold standard here – as for many analytical-chemical questions in clinical chemistry – is triplate-quad mass spectrometry coupled with liquid chromatography (Fig. 5.5).

The molecular ion with mass 2789.4 is detected at triple positive charge at a mass/charge ratio of approx. 930 m/z (Fig. 5.5a). The same ion is found in the isotopically labelled standard at approx. 937 m/z (Fig. 5.5b) and serves as an internal standard which gives the analytical-chemical determination a very high quality (gold standard). The molecular ions together with intensive fragment ions are used for the qualification and quantification of hepcidin in blood serum.

In the future, this method will certainly serve to calibrate and standardize biochemical procedures for the determination of this central regulator of iron metabolism, which can be used in routine clinical practice. However, cost-effective methods based on mass spectrometry will certainly also become established in the future, which will then also be able to determine other isoforms of hepcidin in parallel (Chen et al. 2020).

Mass spectrometry is very efficient, highly specific and universally applicable. It continues to develop into the standard tool of laboratory medicine and will possibly replace immunochemical methods in various fields, including routine clinical chemistry applications.

As can be seen here, the diagnosis of iron deficiency can become very complicated in the presence of certain conditions, and better and better biomarkers are being discovered and more advanced determination technologies are being used in accordance with scientific progress.

Since the conditions in individual patients can be very different, recent publications generally recommend a more personalized approach to the determination of

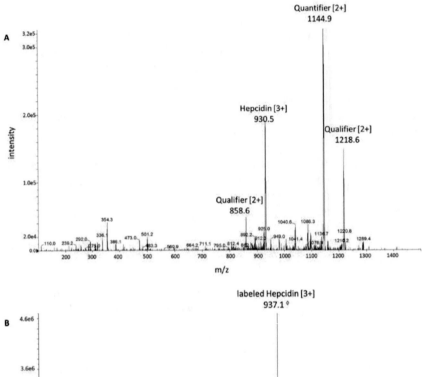

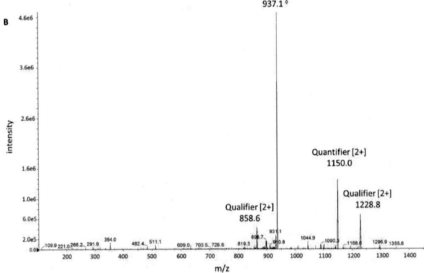

Fig. 5.5 Mass spectrum for the determination of hepcidin-25 with a coupling of liquid chromatography and tripl.quad.mass spectrometry. The mass/charge ratio m/z is plotted against the intensity of the ions. Spectrum (**a**): without isotope labeling. Spectrum (**b**): stable isotope labeling with ^{13}C atoms in a hepcidin molecule (Chen et al. 2020). (Courtesy of John Wiley and Sons)

Mass Spectrometry in Medical Diagnostics

Quads: This is the abbreviation for quadrupoles. They consist of four symmetrically arranged metal rods. By means of special superpositions of DC and AC voltages, they can select and even trap ions with a certain mass-to-charge ratio (trap function). Quadrupoles are descendants of the ion traps for the development of which Wolfgang Paul, professor at the University of Bonn, received the Nobel Prize in Physics in 1989.

Organic tripl.quad.mass spectrometry: This is an arrangement in which three quadrupoles are connected in series. In the first quad, the desired analyte ions, e.g. hepcidin (molecular ion $M = 2789.4$ Da), are selected and only these ions are passed into the second quad. Here they are then fragmented very specifically by impact processes. This fragmentation proceeds according to the laws of chemistry, is quasi a fingerprint of the original molecule and provides information about the structure. The masses of the fragment ions and their intensity are finally determined in the third quad. in combination with a detector. This method is also called tandem mass spectrometry.

Inorganic tripl.quad.mass spectrometry: This is used to determine the elements, e.g. in a blood sample. Almost all elements of the periodic table can be determined, down to the ultratrace range. In front of the inlet to the mass spectrometer there is a hot plasma (ICP: inductively coupled plasma), which fragments all compounds in the sample almost completely. The first quad. then filters out the mass range of interest from the ion stream. In the second quad. interfering ions on the mass traces of the analytes are then removed, e.g. by reaction with hydrogen or methane, and in the third quad. the elements are quantified in combination with the detector. This method, also known as ICP-MS, can also be used in a special mode to determine, for example, nanoparticles of elements directly in whole blood (Witzler et al. 2018).

Couplings with gas chromatography (GC) or liquid chromatography (LC): Both mass spectrometric methods can now be combined with GC or LC. In chromatography, the substances are separated, e.g. in blood serum, and successively fed into mass spectrometry (MS). In inorganic MS, the substances first pass through the hot plasma where they are mainly split into the elements. This LC-ICP-MS combination thus makes it possible to determine the bonding form of elements via their retention time in the LC (species analysis). In organic MS, the intact separated molecules are either ionized by electron impact in the gas phase and then enter the MS (GC-MS) or they are ionized in the liquid phase by electrospray ionization (ESI) after liquid chromatographic separation (usually as HPLC, high performance liquid chromatography) and then brought into the MS. For example, environmental hormones can be determined in blood using this technology in the ultra-trace range (Acir and Günther 2018). Part of the 2002 Nobel Prize in Chemistry was awarded to the American John Fenn for the development of the ESI technique.

(continued)

(continued)

Matrix-assisted laser desorption/ionization mass spectrometry: This revolutionary technique, also known as MALDI-MS for short, has become indispensable in clinical chemistry and also allows the direct analysis of high molecular weight and sensitive molecules such as proteins, polysaccharides or nucleic acids. The analyte is crystallized together with a matrix, e.g. low molecular weight aromatic compounds, and the co-crystallisate is then excited with a UV laser. Through this chemical trick, the molecules are only slightly fragmented and can then be detected with an MS. Japanese Koichi Tanaka was awarded part of the 2002 Nobel Prize in Chemistry for developing this method. SELDI-MS (surface-enhanced laser desorption/ionization mass spectrometry) is a variation of this method. The analytes purified on a chromatographic surface are ionized here after addition of the laser-active matrix as in MALDI and detected via an MS. SELDI-MS is increasingly used, for example, in the search for biomarkers for various diseases.

iron status (keyword: personalized medicine) and present corresponding diagnostic schemes (De Franceschi et al. 2017).

It is also interesting to note that iron deficiency has an influence on the level of the HbA_{1c} value, which is used to control blood glucose over the last 3 months and which can therefore be distorted (English et al. 2015). Also, determining the true iron status in elderly patients is often not straightforward as the usual laboratory parameters may be distorted. Especially in these cases, other possibilities of medical iron diagnostics must therefore be considered (Joosten 2018).

All in all, the field of iron diagnostics is not a simple discipline, and it is therefore naturally not possible to cover all aspects in this chapter. A concise and clear presentation of the subject can be found in the monograph by Hubert Schrezenmeier "Eisenmangelanämie" (Schrezenmeier 2011). Furthermore, up-to-date, practice-oriented information on developments can be obtained from the Portal of Scientific Medicine, where guidelines are issued at regular intervals that also address iron deficiency anemia (Behnisch et al. 2016).

The book series "Laboratory and Diagnosis" already has a tradition of more than 40 years and is designed by renowned representatives of laboratory medicine and clinical chemistry. The latest edition is now available electronically. The contributions available there from various authors are divided into indication, method of determination, examination material, reference range, evaluation, disorders, pathophysiology and literature. In the more than 500 tables, the disease is prefixed, and laboratory diagnostics are described below (Thomas 2020). This new online version is a valuable resource with quickly available valid information.

The strong scientific developments within iron diagnostics, in particular newer examination parameters and techniques of instrumental analytical chemistry, will certainly help to improve differential diagnostics in this field in the future.

References

Acir I, Günther K (2018) Endocrine-disrupting metabolites of alkylphenol ethoxylates – a critical review of analytical methods, environmental occurrences, toxicity, and regulation. Sci Total Environ 635:1530–1546. https://www.sciencedirect.com/science/article/pii/S0048969718312439

Behnisch W, Muckenthaler M, Kulozik A (2016) S1-Leitlinie 025–021: Eisenmangelanämie. AWMF online, das Portal der wissenschaftlichen Medizin. https://www.awmf.org/leitlinien/detail/ll/025-021.html

Chen M, Liu J, Wright B (2020) A sensitive and cost-effective high-performance liquid chromatography/tandem mass spectrometry (multiple reaction monitoring) method for the clinical measurement of serum hepcidin. Rapid Commun Mass Spectrom 34(S1):e8644. https://doi.org/10.1002/rcm.8644

De Franceschi L, Iolascon A, Taher A, Cappellini M (2017) Clinical management of iron deficiency anemia in adults: systemic review on advances in diagnosis and treatment. Eur J Intern Med 42:16–23

English E, Idris I, Smith G, Dhatariya K, Kilpatrick E, John W (2015) The effect of anaemia and abnormalites of erythrocyte indices on HbA1c analysis: a systematic review. Diabetologia 58:1409–1421

Hoppler M, Zeder C, Walczyk T (2009) Quantification of ferritin-bound iron in plant samples by isotope tagging and species-specific isotope dilution mass spectrometry. Anal Chem 81:7368–7372

Joosten E (2018) Iron deficiency anemia in older adults: a review. Geriatr Gerontol Int 18:373–379

Schrezenmeier H (2011) Eisenmangelanämie. Georg Thieme Verlag, Stuttgart

Thomas L (Hrsg.) (2020) Labor und Diagnose. https://www.labor-und-diagnose-2020.de

Witzler M, Küllmer F, Günther K (2018) Validating a single-particle ICP-MS method to measure nanoparticles in human whole blood for nanotoxicology. Anal Lett 51:587–599. https://www.tandfonline.com/doi/full/10.1080/00032719.2017.1327538

The Iron Requirement

6

The daily iron loss via sweat, urine and stool is in the order of 1 mg. In addition, women lose iron through menstruation. During growth and pregnancy, the iron requirement is also increased.

It is currently assumed that on average only 10–15% of the iron in food is available to the body in a mixed diet (bioavailability), and therefore much higher values are given in the recommendations for daily iron intake of the D-A-CH societies and the other international professional societies, which already take this into account. In recent times, however, much more differentiated views on the bioavailability of iron from food have been developed, which in particular take into account the iron status and the specific life situation of the person and at the same time the iron binding form in the consumed foods in a more targeted manner. These interrelationships will be discussed later (Chap. 7).

6.1 International Reference Values

Tables 6.1, 6.2, 6.3, and 6.4 list the reference values for daily iron intake for different groups of people from nine international professional societies, published in German or English. In future, a comparison and scientific comparison of all the reference values available worldwide should certainly be carried out in order to take account of globalization here too.

International Professional Societies of the Iron Reference Values in Tables 6.1, 6.2, 6.3 and 6.4

- D-A-CH: German Nutrition Society, Austrian Nutrition Society, Swiss Nutrition Society
- EFSA: European Food Safety Authority, Parma, Italy, established 2002
- WHO/FAO: World Health Organization/Food and Agriculture Organization of the United Nations
- DoH: Department of Health, United Kingdom
- NAM: National Academy of Medicine, Washington DC, USA
- NHMRC: National Health and Medical Research Council, Australia and New Zealand
- Health Council: The Hague, Netherlands
- Nordic Council: Copenhagen, Denmark
- FSA: Food Safety Authority of Ireland
- A compilation of the various reference values can also be found in Schelest 2019.

Table 6.1 Reference values for adults for daily iron intake from international professional societies

Institution	Range (years)	Reference value Women (mg Fe/d)	Reference value Men (mg Fe/d)
D-A-CH	19–51	15	10
	>51	10	10
EFSA	>18	16	11
WHO/FAO[a]	>18	19.6/24.5/29.4/58.8[b] 7.5/9.4/11.3/22.6[c]	9.1/11.4/13.7/27.4
DoH (UK)	19–49	14.8	8.7
	>50	8.7	8.7
NAM (USA, Canada)	19–50	18	8
	51– >70	8	8
NHMRC (AUS, NZL)	19–50	18	8
	52– >70	8	8
Health Council (NL)	>18	16(b)11(c)	11
Nordic Councils of Ministers	18–30	15	9
	31–60	15(b)/9(c)	9
	61– >75	9	9
FSA of Ireland	19–54	14	10
	>55	9	
	19– >65		

D-A-CH 2018; EFSA 2015; WHO/FAO 2004; DoH 2009; NAM 2001; NHMRC 2006; Health Council 1992; Nordic Council of Ministers 2014; FSA 1999
[a]Reference values for bioavailabilities 15%, 12%, 10% und 5%
[b]Premenopause
[c]Postmenopause

Table 6.2 Reference values for pregnant and lactating women for daily iron intake from international professional societies

Institution	Pregnant (mg Fe/d)	Breastfeeding (mg Fe/d)
D-A-CH	30	20
EFSA	16	16
WHO/FAO[a]	–	10/12.5/15/30
DoH (UK)	14.8	14.8
NAM (USA, Canada)	27	10
NHMRC (AUS, NZL)	27	10
Health Council (NL)	11	20
Nordic Councils of Ministers	–	15
FSA of Ireland	15	15

D-A-CH 2018; EFSA 2015; WHO/FAO 2004; DoH 2009; NAM 2001; NHMRC 2006; Health Council 1992; Nordic Council of Ministers 2014; FSA 1999
[a]Reference values for bioavailabilities 15%, 12%, 10% und 5%

When considering the values of the various professional societies for adults, a heterogeneous picture emerges in some cases, with the WHO/FAO data given as a function of the bioavailability of iron (Table 6.1). The values of the eight other societies range between 14 and 18 mg iron/day for women of childbearing age and between 8 and 11 mg iron/day for men. The D-A-CH and EFSA values differ only slightly.

The reference values for pregnant women and breastfeeding women, on the other hand, differ considerably in some cases. For example, the D-A-CH recommendation for pregnant women of 30 mg iron per day is almost twice as high as the EFSA value of 16 mg/day. The NAM and NHMRC data for the USA, Australia and New Zealand are in a similar range to the D-A-CH with 27 mg iron/day, and the lowest reference value is found with 11 mg/day by the Dutch professional association, which is only about one third of the D-A-CH data. The WHO/FAO does not currently provide any information for pregnant women (Table 6.2).

The reference values for breastfeeding women do not differ as much as those for pregnant women, but here too there are differences of up to 100%. For example, for the USA, Canada, Australia and New Zealand, quantities of 10 mg Fe/day can be found, and for Germany, Austria and Switzerland, 20 mg Fe/day. The WHO/FAO figures are again based on different bioavailabilities of the element of 15%, 12%, 10% and 5% (Table 6.2).

The reference values for children and adolescents are partly divided according to different age groups by the various professional societies, and the WHO/FHO figures are again based on different bioavailabilities. In the highest age groups, the values range from 13 to 15 mg Fe/day for females and between 11 and 14 mg Fe/day for males (Table 6.3). Larger deviations, as in the case of pregnant and breastfeeding women (cf. Table 6.2), are not found here for any age group. In the case of the reference values for infants, the information provided by the individual professional societies varies in some cases. In the 6–12 month age group, they range from 7 to 11 mg Fe/day, which is a difference of up to 75%. For infants aged up to

Table 6.3 Reference values for children and adolescents for daily iron intake from international professional societies

Institution	Range (years)	Reference value female (mg Fe/d)	Reference value masculine (mg Fe/d)
D-A-CH	1–7	8	8
	7–10	10	10
	10–19	15	12
EFSA	1–6	7	7
	7–11	11	11
	12–17	13	11
WHO/FAO[a]	1–3	3.9/4.8/5.8/11.6	3.9/4.8/5.8/11.6
	4–6	4.2/5.3/6.3/12.6	4.2/5.3/6.3/12.6
	7–10	5.9/7.4/8.9/17.8	5.9/7.4/8.9/17.8
	11–14	9.3/11.7/14.0/28.0[b]	9.7/12.2/14.6/29.2
	15–17	21.8/27.7/32.7/65.4[c] 20.7/25.8/31.0/62.0	12.5/15.7/18.8/37.6
DoH (UK)	1–3	6.9	6.9
	4–6	6.1	6.1
	7–10	8.7	8.7
	11–18	14.8	11.3
NAM (USA, Canada)	1–3	7	7
	4–8	10	10
	9–13	8	8
	14–18	15	11
NHMRC (AUS, NZL)	1–3	9	9
	4–8	10	10
	9–13	8	8
	14–18	15	11
Health Council (NL)	1–7	7	7
	7–10	8	8
	10–13	11	10
	13–16	15	12
	16–19	15	14
Nordic Councils of Ministers	1–5	8	8
	6–9	9	9
	10–13	11	11
	14–17	15	11
FSA of Ireland	1–3	8	8
	4–6	9	9
	7–10	10	10
	11–14	14	13
	15–18	14	14

D-A-CH 2018; EFSA 2015; WHO/FAO 2004; DoH 2009; NAM 2001; NHMRC 2006; Health Council 1992; Nordic Council of Ministers 2014; FSA 1999
[a]Reference values for bioavailabilities 15%, 12%, 10%, 5%
[b]Premenarche
[c]Postmenarche

Table 6.4 Reference values for infants for daily iron intake from international professional societies

Institution	Range (month)	Reference value (mg Fe/d)
D-A-CH	0–4	0.5
	4–12	8
EFSA	0–6	n. i.
	7–12	11
WHO/FAO[a]	0–6	n. i.
	6–12	6.2/7.7/9.3/18.6
DoH (UK)	0–3	1.7
	4–6	4.3
	7–12	7.8
NAM (USA, Canada)	0–6	0.27
	7–12	11
NHMRC (AUS, NZL)	0–6	0.2
	7–12	11
Health Council (NL)	0–6	5
	6–12	7
Nordic Councils of Ministers	0–6	n. i.
	6–11	8
FSA of Ireland	0–3	1.7
	4–6	4.3
	7–12	7.8

D-A-CH 2018; EFSA 2015; WHO/FAO 2004; DoH 2009; NAM 2001; NHMRC 2006; Health Council 1992; Nordic Council of Ministers 2014; FSA 1999
[a]Reference values for bioavailabilities of 15%, 12%, 10% und 5%
n.i. no information

4–6 months, reference values of 0.2–5 mg Fe/day are given, indicating a considerable difference (Table 6.4).

The comparison of the reference values of international professional societies for daily iron intake shows quite good agreement for some groups of people, while in some cases large differences can be found in the data. In the future, further processing and scientific penetration of the topic will certainly be necessary in order to present reliable values with an even more precise derivation in a more internationally and globally oriented nutrition and food science.

In the context of these activities, it is also very important to integrate and review research results from various fields, in particular bioavailability (Dainty et al. 2014), into these considerations, or to critically evaluate and question statements about iron losses (Green et al. 1968; Hunt et al. 2009), which have previously played a role in the derivation of reference values.

Of the nine institutions and professional societies considered, four use the older study by Green et al. to calculate basal losses, while EFSA refers to the research by Hunt et al. from 2009.

EFSA used models for the determination of bioavailability for the derivation of its reference values, which use iron absorption as a function of iron status. This is determined via the serum ferritin value and is, however, also dependent on inflammation parameters. Other of the institutions listed determine bioavailabilities via food composition.

A combination of both approaches would certainly be indicated for the future, since both influencing factors can have a decisive effect on the absorption of iron in the organism. For this purpose, further scientific investigations in both fields mentioned are necessary. Extended stable-isotope studies must put the absorption models on a better footing, and a comprehensive species analysis of iron must elucidate the binding forms of the element in the main food groups in terms of structural chemistry (cf. Chap. 7). Unfortunately, the complexes that iron forms with food constituents are still not known in many cases.

Furthermore, more attention should also be paid to the influence of hepcidin. Examples include the fact that a decrease in hepcidin levels leads to an increase in the absorption of iron from food (Fisher and Nemeth 2017), and the maternal hepcidin concentration also appears to have an influence on the iron absorption of the fetus (Koenig et al. 2014).

These are only a few examples of results that should be included in a future evaluation of different statements by the various professional societies. A clear, comprehensible derivation of the reference values must be in the foreground.

The summary of the derivation of EFSA's 2015 iron reference values is presented below.

Results of the Derivation of EFSA Dietary Reference Values (DRVs) for Iron

The average requirement (AR) and population reference intake (PRI) were determined. In adults, whole body iron losses were modeled using data from US adults. Predicted absorption values at a serum ferritin concentration of 30 µg/l of 16% for men and 18% for women were used to account for physiologic conditions in iron intake. Iron absorption rates depend on the iron status of the organism, as determined by serum ferritin levels. In men, median whole-body iron losses are 0.95 mg/day, and the AR is then 6 mg/day at 16% absorption. The PRI, calculated as the requirement at the 97.5th percentile, is 11 mg/day. The same DRVs are suggested for postmenopausal women as for men. In premenopausal women, additional iron is lost through menstruation, but because losses are highly skewed, the EFSA Panel set a PRI of 16 mg/day to meet the needs of 95% of the population. For infants and children, requirements were factorially calculated taking into account growth requirements, replacement of losses and percentage dietary iron intake (10% up to 11 years and 16% thereafter). PRIs were estimated with a coefficient of variation of 20%. They are 11 mg/day for infants (7–11 months), 7 mg/day for children aged 1–6 years, and 11 mg/day for children aged 7–11 years and boys aged 12–17 years. For girls aged 12–17 years, the PRI of 13 mg/day is the midpoint of the calculated nutritional requirements of 97.5% of girls and the PRI for premenopausal women; this approach allows for the large uncertainties in the rate and timing of pubertal growth and menarche. For pregnant and lactating women, for whom iron stores and enhanced absorption were assumed to provide sufficient supplemental iron, the DRVs are the same as for premenopausal women (EFSA 2015).

In order to demonstrate how the reference values can be derived, an example will now be presented explicitly with quantitative data and calculations. The EFSA iron reference value of 16 mg/day for pregnant women was selected because it differs significantly from the corresponding D-A-CH value of 30 mg/day (cf. Table 6.2).

The following reflects EFSA's derivations for pregnant women (EFSA 2015): for a single pregnancy of an average adult woman, 835 mg of iron is required over the entire period. This was factorially calculated by the Panel as follows: Total losses (faecal, via urine and dermal) 300 mg, 270 mg is required for the newborn (Bothwell 2000; Milman 2006), 90 mg is required for the placenta and umbilical cord (Bothwell 2000; Milman 2006), and 175 mg of iron is lost through blood loss at birth (mean of Bothwell 2000 and Milman 2006 values).

The calculated loss of 835 mg of iron throughout pregnancy is contrasted by the EFSA Panel with two different approaches to calculating the absolute absorption rates of iron throughout this period: Isotopic studies by Barrett et al. (1994) and use of the model by Dainty et al. (2014).

Isotopic study by Barret et al.:

During pregnancy, there is an expansion of plasma and blood volume and red blood cell mass that begins at 6–8 weeks and peaks at 28–34 weeks of gestation. The dilution effect of this expansion induces a fall in serum ferritin concentration to about 15 µg/l, although the link to systemic iron stores is not lost (Blackburn 2012). The increased demand for iron is also met by an increase in the efficiency of iron absorption (Bothwell et al. 1979; Hallberg and Hultén 1996).

The absorption rates of dietary iron during pregnancy were studied using isotopic labelling in a group of 12 women on a diet containing 9 mg of non-haem iron daily (Barrett et al. 1994). An increase in iron absorption was found in the three trimesters of pregnancy. In parallel, serum ferritin concentrations decreased, reflecting the expansion of plasma volume. The increase in iron absorption in healthy women eating a mixed diet may compensate for the increased requirement in later pregnancy, as has also been shown in other isotope studies in pregnant women (Whittaker et al. 1991, 2001).

The amount of iron absorbed can be predicted using data from an isotopic study (Barrett et al. 1994): under the conservative assumption that the same percentage iron absorption observed at 12 weeks' gestation applies to the period of gestational weeks 0–23, and that the percentage iron absorption observed at 24 weeks' gestation applies to the period of gestational weeks 24–35, and that the percentage iron absorption observed at 36 weeks' gestation applies to the period of gestational weeks 24–35. The percentage iron absorption observed at 36 weeks of gestation applies to the period 36–40 weeks of gestation.

The amount of non-haem iron absorbed (mg/day) was calculated assuming dietary non-haem iron of 9 mg/day and 4 mg haem iron/day from meat throughout pregnancy. As there is no evidence of an increase in haem iron absorption during pregnancy (Young et al. 2010), it is assumed to be 25% at all stages of pregnancy. However, the Panel considers that this may be too low a value because of insufficient data on the efficiency of haem iron absorption throughout pregnancy.

The following data for iron absorption during pregnancy, calculated on the basis of data from Barrett et al. (1994), were used to further derive the values.

Absorbed non-heme iron in mg/day from a daily administration of 9 mg:

Day 1–161 (161 days total): (7.2% absorption) 0.65 mg/day, total: 104.65 mg
  iron intake.
Day 162–245 (84 days total): (36.3% absorption) 3.27 mg/day, total: 274.68 mg
  iron intake.
Day 246–280 (35 days total): (66.1% absorption) 5.95 mg/day, Total: 208.25 mg
  iron intake.
Total intake of non-heme iron throughout pregnancy: 587.58 mg.

Absorbed heme iron in mg/day from a daily administration of 4 mg:

Absorption was assumed to be 25%, unchanged throughout pregnancy for a total of
  280 days: 280 mg.
Absorbed non-heme iron and heme iron over the entire period of pregnancy:
  587.58 mg + 280 mg = 867.58 mg iron.

As the amount of iron needed in pregnancy is about 835 mg (see above), no additional dietary iron is needed. The Panel notes that the experimental studies from
which the derivations result could still be improved in design to increase the informative value (EFSA 2015).

Model of Dainty et al.:

Using factorial estimation, an alternative method for calculating bioavailability
factors to derive DRVs was developed by Dainty et al. (2014). This was done using
data collected for the NDNS (National Diet and Nutrition Survey). These are from
a nationally representative sample of adults living in the UK and consuming a mixed
Western diet. They include serum ferritin concentration and total (haem and non-
haem) iron intake determined from a 7-day food diary. The acute phase marker
antichymotrypsin was determined to ensure that the data used were from individuals
who were free of inflammation. The NDNS sample included 495 men and 378 premenopausal women.

The model can be used to predict iron absorption at any serum ferritin concentration. For example, at a serum ferritin concentration of 60 µg/l, iron absorption would
be 11% in both men and premenopausal women. Serum ferritin concentration of
30 µg/l is correlated with iron absorption of 18% in women and 16% in men.

Assuming serum ferritin concentrations of 30 µg/L (early pregnancy, up to
23 weeks), which is associated with an iron absorption efficiency of 18%, and
15 µg/l (late pregnancy, from 24 weeks to the end of pregnancy), which is associated
with an iron absorption efficiency of 31%, the amount of iron absorbed from a
mixed diet can be calculated.

At a serum ferritin concentration of 30 µg/l, total food intake must be 4639 mg
(at 18% absorption) to provide 835 mg of absorbed iron (i.e., the total amount of
iron needed for pregnancy), which is equivalent to 16.6 mg/day for 280 days of the
total period. At a serum ferritin concentration of 15 µg/l, absorption is 31%, and
total dietary intake must be 2694 mg to provide the required amount of 835 mg,
which is equivalent to 9.6 mg/day.

In practice, serum ferritin concentration will gradually decrease as pregnancy progresses, and taking the mean of these two estimates, the average dietary intake to provide the required amount of iron would be 13.1 mg/day. Assuming a CV (coefficient of variation) of 20% to account for large interindividual variations in iron requirements in pregnant women, this would correspond to a theoretical PRI of 18.3 mg/day. This theoretical calculation is an alternative approach to using the percentage iron absorption values derived from the isotopic studies and is based solely on the relationship between serum ferritin concentration and iron absorption efficiency.

The Panel notes in summary that the conclusion from these two different approaches is similar in that there is no need for additional dietary iron during pregnancy, provided that adequate iron stores were available at conception. This is a consequence of the increasing efficiency of iron absorption during pregnancy. However, the Panel also notes that the Dainty et al. (2014) model has not been validated for pregnant women. Furthermore, it does not take into account adaptive changes in absorption efficiency that occur during pregnancy (EFSA 2015).

As no additional dietary iron is required for pregnant women on the basis of these two derivations, the same reference value of 16 mg/day results for this group (cf. Table 6.2) as for women aged >18 years, for whom 16 mg/day is also estimated by EFSA (cf. Table 6.1).

In summary, it can be said that the reference values for the daily intake of iron in the recommendations of the individual professional societies are partly similar, but there are also areas where the differences are very large (Tables 6.1, 6.2, 6.3, and 6.4). In the future, there is an urgent need for internationally coordinated nutritional standards and reference values from the individual professional societies, which can make a good contribution to assessing the quality of the world's food supply in the course of further globalization (cf. Sect. 4.1).

Important Aspects for the Future Development of the Reference Values
Simultaneous consideration of the three factors that have an influence on iron absorption: Homeostasis, iron species present and accompanying substances in the food (cf. Fig. 7.1).

New isotope studies and derived theoretical models for describing the resorption rate as a function of iron status need to be more closely adapted to real-life conditions. For example, it is important to also work with the labelled iron species that frequently occur in foods, and not only with simple iron salts.

The rough subdivision into heme iron and non-heme iron is no longer sufficient according to current knowledge, and ferritin iron must be considered separately; this should also be discussed for other important iron species (cf. Fig. 7.4).

The influence of food additives on the absorption of iron should be re-evaluated on the basis of the current literature, as the studies are partly contradictory. Here, a close examination of the validity of the results in the corresponding publications is indicated.

A specification of reference values for iron depending on other parameters such as the present iron species or the presence of certain food ingredients, as in the new zinc reference values (Haase et al. 2020), should be discussed intensively.

6.2   Nutrient Reference Value for the Daily Intake of Iron

In the European Food Information Regulation (EU 2011), nutrient reference values, often abbreviated to NRV (Nutrient Reference Value), were defined in connection with the standardization of Europe-wide regulations. The NRVs indicate the amounts of trace elements, minerals and vitamins that an average adult should consume daily to cover his or her needs. They do not take into account gender or age and are to be seen as uniform reference values to which the daily intake can be generally validly related. For iron, the NRV is 14 mg.

The NRV is also important for information on food packaging. The percentage information that can be found on the nutritional value tables, if iron has been specified there, is based on this quantity. In parentheses it is stated how much % iron of these 14 mg is incooperated when consuming a portion of 100 g of the food.

In Chap. 8 it is shown that this 14 mg can be achieved very well by a well composed combination of different foods and dishes. For the sake of clarity, the intake of iron is given there as a percentage of 14 mg in two examples and the calculation is presented explicitly (cf. Figs. 8.28 and 8.30). Thus, by simply adding up the percentages, the daily iron intake of the patients can be calculated in a rollover. If the sum results in 100%, the daily optimal iron intake has been reached.

In Chap. 8, however, it is also described that a certain combination of foods, which can certainly be described as healthy in the general sense, may mean that one hardly absorbs any iron and thus unconsciously slips into a severe iron deficiency situation without suspecting it. This then leads to the fact that numerous complaints announce themselves or develop and one is completely helpless, since the patient nourishes himself nevertheless so healthy. A list of foods with a low iron content of <0.5 mg/100 g can be found in the iron negative list (Sect. 8.2).

6.3   Upper Limit of Iron Intake

Essential trace elements such as iron, zinc or selenium show a characteristic behavior in the human organism, which is shown in Fig. 6.1. It is always important to stay within the safe intake range. Too much is just as harmful as too little. This is also the quintessence of the well-known saying of the famous physician and philosopher Theophrastus Bombast von Hohenheim, known as Paracelsus, which dates back to the sixteenth century: "All things are poison and nothing is without poison; it is only the dose that makes a thing not a poison."

If the element is missing in the human body, or if there is much too much, this can lead to death in extreme cases. If there is a slight deficiency or a not too strong surplus, this has negative effects on the organism, which can be more or less strong. If the dose of the element is within a certain range, the body is optimally supplied with the essential trace element and it can function at its best. In Fig. 6.1, this range is labeled Health and forms the plateau in the curve shown.

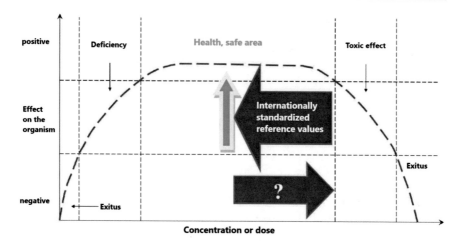

Fig. 6.1 Dose-response diagram for a typical essential element. The onset of the toxic effect (right-hand side of the diagram) is fairly well known for many trace elements, but is still relatively unclear for iron (red arrow to the right). In the future, globally valid uniform reference values should be defined

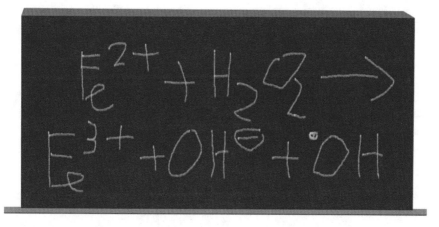

Fig. 6.2 Blackboard picture from a lecture at the University of Bonn: Formation of an OH radical, the point at the OH symbolizes the lone electron (so-called Fenton-Haber-Weiss reaction)

This plateau is also referred to as the therapeutic width. It is strongly dependent on which trace element I am looking at and in a way represents the ratio of too little and too much. If the supply of the trace element is within this plateau, the body is optimally supplied.

For iron, there is currently a great deal of scientific uncertainty about a well-defined upper limit of iron intake, and knowledge of this is very important to avoid adverse health effects (Fig. 6.1).

In the organism, hydrogen peroxide (H_2O_2) is formed during various metabolic processes and is normally destroyed by enzymes (catalase), as it could otherwise trigger dangerous reactions. In the case of excess iron, however, this compound can be converted into hydroxyl radicals in a special reaction (Fig. 6.2).

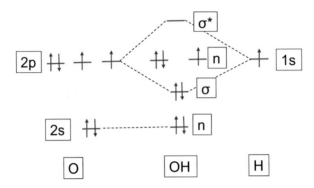

Fig. 6.3 LCAO-MO diagram of the hydroxyl radical. LCAO: Linear Combination of Atomic Orbitals. MO: Molecular Orbital. Diagram without 1 s level of oxygen. The bond order is 1, and the free electron is in the higher energy non-bonding MO. Non-bonding orbital: n, binding sigma orbital: σ, antibonding sigma orbital: σ*

The OH or hydroxyl radicals have a very high oxidation potential, which can be seen very well in the scheme of the molecular orbital theory, where an unpaired electron is located in the higher energy non-bonding molecular orbital (red). Hydroxyl radicals are therefore extremely reactive (Fig. 6.3).

Hydroxyl or OH radicals play a major role in metabolism and are also held responsible, among other things, for the development of numerous neurodegenerative diseases such as Alzheimer's disease or Parkinson's disease. They are formed by the reaction of Fe^{2+} with H_2O_2 and belong to the group of reactive oxygen species (ROS). Due to the high electrode potential of the system $\dot{O}H, H^+/H_2O$ of +2.31 V (pH 7.0), the hydroxyl radical reacts with many biomolecules under hydrogen abstraction and by hydroxyl addition. The reaction is diffusion controlled with second order kinetics and a rate constant of $k \approx (0.5–2) \times 10^{10} \ M^{-1} \ s^{-1}$. The hydroxyl radical is therefore one of the most chemically active ROS and can therefore cause a wide variety of oxidative damage to nucleic acids, membranes, enzyme systems or whole cells (oxidative stress) (Zhao 2019). ROS also play an important role in ferroptosis, iron-dependent cell death (see Fig. 6.4). It is also interesting to note that the hydroxyl radical, which has a very central function in physiological chemistry, also plays a major role in the degradation of many trace gases in the atmosphere ("atmospheric detergent") and it was even possible to detect it in interstellar space by spectroscopic methods as early as the 1960s by scientists at the Massachusetts Institute of Technology, USA (Weinreb et al. 1963).

(continued)

(continued)

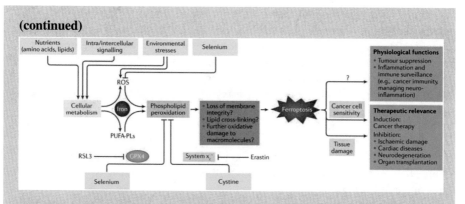

Fig. 6.4 Overview of ferroptosis. The mechanism of cell death is initiated by the formation of reactive oxygen species (ROS) through the involvement of iron. Peroxidation of phospholipids containing polyunsaturated fatty acids occurs, leading to membrane damage. An important regulator of ferroptosis is selenium. GPX4: glutathione peroxidase (Jiang et al. 2021). (Courtesy of Springer Nature Limited)

The Fenton-Haber-Weiss reaction and the resulting hydroxyl radicals can of course cause numerous damages in the body. This can especially take place if free iron is present in the cells, which is not bound to transferrin, ferritin, haemoglobin or enzymes and thus good chemical reaction possibilities are given. In somewhat simplified terms, it can be said that these processes take place in different ways in the various cells and organs of the body and that it is therefore not yet possible to give an exact answer to the questions "When is there too much iron, when does damage occur?." This is an important difference to many other trace elements or vitamins.

A review article on this subject sheds light on the difficulties of setting a defined upper limit for iron intake from numerous different scientific perspectives, as the different toxic mechanisms of action of excess iron are very complex. It also discusses the risks of iron overload. The article cites two studies that indicate an upper limit of total daily intake (i.e., counting dietary iron) of about 40 mg. A clearly defined upper limit is not given, and this question remains difficult to answer (Schümann et al. 2014).

The Federal Institute for Risk Assessment (BfR) in Berlin is even more critical of additional iron intake without a doctor's prescription. The BfR explains that a persistent oversupply of iron increases the risks for the development of cardiovascular diseases, diabetes and cancer. However, no dose-response relationship can be derived between iron and these diseases, so there are considerable uncertainties in setting a maximum amount. Furthermore, the BfR is against the consumption of breakfast cereals (cereals and cereal products) fortified with iron. Iron-containing food supplements should only be taken in cases of proven iron deficiency and after consultation with a doctor (BfR 2009 and 2013).

All in all, the reference values for the daily intake of iron for the individual age groups and groups of people must be better coordinated between the various professional societies and organizations in the future. In addition, a clear indication of the maximum acceptable amount of iron supplementation would be important in order to obtain more certainty in this direction as well.

Reactive oxygen species and thus hydroxyl radicals, which are formed by reaction with iron ions, also play the central role in ferroptosis, which was discovered in animal cells a few years ago (Dixon et al. 2012). The biochemical mechanism of iron-dependent cell death is also found in plants (Distefano et al. 2017), and aging processes are also associated with it (Jenkins et al. 2020). Therefore, this new form of apoptosis under regulation of iron ions will be briefly presented in its basic features in the following (Fig. 6.4).

An essential part of the normal development and maintenance of the whole life of an organism is cell death. It can occur through caspase-mediated apoptosis or through non-apoptotic processes. In 2002, the Nobel Prize in Medicine was awarded to Sydney Brenner, John Sulston and Robert Horvitz for their fundamental work on apoptosis, also known as programmed cell death. Their investigations on the model organism C. Elegans proved for the first time that certain cells die in a planned manner and that this process is genetically controlled. These discoveries are very important for medical research, since apoptosis is also genetically determined in humans and mutations of the corresponding genes can trigger diseases.

Ferroptosis is a non-apoptotic cell death. It is an oxidative and, in particular, iron-dependent process that results from the formation of lipid-reactive oxygen species. Ferroptosis does not require a specific effector protein, and therefore this process is thus in contrast to apoptosis and other forms of non-apoptotic cell death. Ferroptosis occurs through oxidative destruction of membrane lipids, and iron ions are at the center of this biochemical mechanism (Fig. 6.4).

Extracellular iron is bound by transferrin, transported into the cell by the transferrin receptor and finally released in the lysosome from the transferrin/transferrin receptor complex into the cytoplasm. There, hydroxyl radicals can then be generated via the Fenton-Haber-Weiss reaction, which then further react with polyunsaturated fatty acid residues in membrane phospholipids. The resulting lipid radicals react with oxygen to form highly reactive lipid peroxyl radicals, which then react with further phospholipids, leading to the formation of further lipid peroxides and veritable membrane damage or overt membrane permeabilization.

Highly reactive lipid peroxides are usually reduced to non-reactive lipid alcohols in the organism by the activity of glutathione hydroperoxidase 4 (GPX4). GPX4 is an enzyme that contains selenium in the form of selenocysteine and requires the reduced form of glutathione (GSH) for its catalytic activity.

The mechanism of ferroptosis clearly shows that iron can intervene very centrally in essential biochemical processes in the cell cycle, and therefore these findings should be taken into account in future considerations of the evaluation of reference values and in setting the upper limit of daily iron intake.

References

Barrett J, Whittaker P, Williams J, Lind T (1994) Absorption of non-haem iron from food during normal pregnancy. BMJ 309:79–82

BfR, Stellungnahme des Bundesinstituts für Risikobewertung (BfR), Nr. 016/2009 vom 2. März 2009, ergänzt am 21. Januar 2013, Verwendung von Eisen in Nahrungsergänzungsmitteln und zur Anreicherung von Lebensmitteln

Blackburn S (2012) Maternal, fetal and neonatal physiology: a clinical perspective, 4th edn. Elsevier Saunder, Philadelphia

Bothwell T (2000) Iron requirements in pregnancy and strategies to meet them. Am J Clin Nutr 72:257S–264S

Bothwell T, Charlton R, Cook J, Finch C (1979) Iron metabolism in man. Blackwell Scientific, Oxford, UK

D-A-CH, Deutsche Gesellschaft für Ernährung, Österreichische Gesellschaft für Ernährung, Schweizerische Gesellschaft für Ernährung (eds) (2018) Referenzwerte für die Nährstoffzufuhr, 2. Auflage, 4. aktualisierte Ausgabe, 3. Ergänzungslieferung 2018, Bonn

Dainty J, Berry R, Lynch S, Hervey L, Fairweather-Tait S (2014) Estimation of dietary iron bioavailability from food iron intake and iron status. PLoS One 9(10):e111824. https://doi.org/10.1371/journal.pone.0111824

Distefano A, Martin M, Cordoba J, Bellido A, D'Ippolito S, Colman S, Soto D, Roldan J, Bartoli C, Zabaleta E, Fiol D, Stockwell B, Dixon S, Pagnussat G (2017) Heat stress induces ferroptosis-like cell death in plants. J Cell Bio 216:463–476

Dixon S, Lemberg K, Lamprecht M, Skouta R, Zaitsev E, Gleason C, Patel D, Bauer A, Cantley A, Yang W, Morrison B III, Stockwell B (2012) Ferroptosis: an iron-dependent form of nonapoptotic cell death. Cell 149:1060–1072

DoH, Department of Health (Hrsg) (2009) Dietary reference values for food energy and nutrients for the United Kingdom: report of the panel on dietary reference values of the committee on medical aspects of food policy. Her Majesty's Stationery Office, 19th edn

EFSA Panel on Dietetic Products, Nutrition and Allergies (2015) Scientific opinion on dietary reference values for iron. EFSA J 13(10):4254. https://doi.org/10.2903/j.efsa.2015.4254

EU (2011) Verordnung Nr. 1169/2011 des Europäischen Parlaments und des Rates vom 25. Oktober 2011

Fisher A, Nemeth E (2017) Iron homeostasis during pregnancy. Am J Clin Nutr 106(Suppl 6):1567S–1574S

FSA, Food Safety Authority of Ireland (Hrsg) (1999) Recommended dietary allowances for Ireland: Nutrition Food Safety Authority of Ireland

Green R, Charlton R, Seftel H, Bothwell T, Mayet F, Adams B, Finch C, Layrisse M (1968) Body iron excretion in man: a collaborative study. Am J Med 45:336–353

Haase H, Ellinger S, Linseisen J, Neuhäuser-Berthold M, Richter M, on behalf of the German Nutrition Society (DGE) (2020) Revised D-A-CH-reference values for the intake of zinc. J Trace Elements Med Bio 61:126536

Hallberg L, Hultén L (1996) Iron requirements, iron balance and iron deficiency in menstruating and pregnant women. In: Asp NG (ed) Iron nutrition in health and disease. George Libbey, London, pp 165–182

Health Council, Voorlichtingsbureau voor de Voeding (Hrsg) (1992) Nederlandse Voedingsnormen, Den Haag

Hunt J, Zito C, Johnson L (2009) Body iron excretion by healthy men and woman. Am J Clin Nutr 89:1792–1798

Jenkins N, James S, Salim A, Sumardy F, Speed T, Conrad M, Richardson D, Bush A, McColl G (2020) Changes in ferrous iron and glutathione promote ferroptosis and frailty in aging Caenorhabditis elegans. eLife 9:e56580. https://doi.org/10.7554/eLife.56580

Jiang X, Stockwell B, Conrad M (2021) Ferroptosis: mechanisms, biology and role in disease. Nat Rev Mol Cell Biol 22:266–282. https://doi.org/10.1038/s41580-020-00324-8

Koenig M, Tussing-Humphreys L, Day J, Cadwell B, Nemeth E (2014) Hepcidin and iron homeostasis during pregnancy. Nutrients 6:3062–3083

Milman N (2006) Iron and pregnancy – a delicate balance. Ann Hematol 85:559–565

NAM, Institute of Medicine (ed) (2001) Dietary reference intakes for vitamin a, vitamin K, arse-
nic, boron, chromium, copper, iodine, iron, manganese, molybdenum, nickel, silicon, vana-
dium, and zinc. National Academies Press, Washington, DC

NHMRC, National Health and Medical Research Council (Hrsg) (2006) Nutrient reference values
for Australia and New Zealand: Including recommended dietary intakes, Canberra

Nordic Council of Ministers (Hrsg) (2014) Nordic Nutrition Recommendations 2012. Integrating
nutrition and physical activity, 5th edn. Kopenhagen

Schelest A (2019) Empfehlungen für die nutritive Zufuhr von Eisen – vergleichende Bewertung
internationaler Referenzwerte. Universität Bonn, Master-Arbeit

Schümann K, Ettle T, Szegner B, Elsenhans B, Solomons N (2014) Risiken und Nutzen der
Eisensupplementation: Empfehlungen zur Eisenaufnahme kritisch betrachtet. Pers Med 2:19–39

Weinreb S, Barrett A, Meeks M, Henry J (1963) Radio observations of OH in the interstellar
medium. Nature 200:829–831

Whittaker P, Lind T, Williams J (1991) Iron absorption during normal human pregnancy: a study
using stable isotopes. Br J Nutr 65:457–463

Whittaker P, Barrett J, Lind T (2001) The erythrocyte incorporation of absorbed non-haem iron in
pregnant women. Br J Nutr 86:323–329

WHO, World Health Organization, Food and Agriculture Organization (Hrsg) (2004) Vitamin and
mineral requirements in human nutrition, 2nd edn. Bangkok

Young M, Griffin I, Pressman E, McIntyre A, Cooper E, McNanley T, Harris Z, Westerman M,
O'Brien K (2010) Utilization of iron from an animal-based iron source is greater than that of
ferrous sulfate in pregnant and nonpregnant women. J Nutri 140:2162–2166

Zhao Z (2019) Iron and oxidizing species in oxidative stress and Alzheimer's disease. Aging
Med 2:82–87

Bioavailability

7

In addition to the iron content of the foods consumed, a very important aspect of iron deficiency nutrition is absorption in the organism, i.e. the bioavailability of the essential trace element. Bioavailability is a very complex parameter that is subject to many influences (Fig. 7.1).

On the one hand, absorption is a function of the iron binding forms (species) in the food, but other food components can also have a significant influence on bioavailability. Furthermore, the regulation of iron absorption via homeostasis is an important point, which means that the supply state of the body with iron can have an effect on the bioavailability of the element (Fig. 7.2). These relationships illustrate that the assessment of absorption and thus also the exact definition and derivation of reference values is a major scientific task that has so far only been partially solved. Some aspects of this have already been discussed in Chap. 6, and it is very important to consider all three main influencing factors simultaneously, which has not yet been done in the derivation of the reference values of all professional societies due to the complex and still insufficient data situation.

To answer the questions about bioavailabilities, the constructive interaction of a wide range of specialist disciplines is required. Experts from analytical chemistry and food chemistry are investigating the chemical composition and stability of the various iron complexes in food. In addition, other food constituents must be identified here that can influence the resorption of iron ions by recomplexation. Finally, the determination of other, in particular bivalent cations in foods is also important, since they, such as zinc, compete for binding sites at the divalent metal transporter 1 (DMT-1, cf. Chap. 3) and can thus negatively influence iron resorption.

The questions from the perspective of nutritional science and nutritional medicine deal with the quantitative aspects of iron absorption with regard to the entire organism. Parameters of iron biochemistry in the organism for which target values can be defined are, for example, the Hb and ferritin values or the transferrin saturation with simultaneous determination of the inflammation marker CRP (cf. Chap. 5). To complete the overall picture, the variability of the blood levels of soluble transferrin

K. Günther, *Diet for Iron Deficiency*, https://doi.org/10.1007/978-3-662-65608-2_7

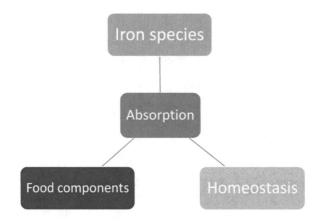

Fig. 7.1 The triad of bioavailability in elements. Three influencing factors determine the absorption of iron in the organism. The joint consideration of all three areas is generally very important when deriving reference values for minerals and trace elements. When establishing reference values or in bioavailability studies for organic substances and micronutrients, biochemical transformation processes (e.g. also in the microbiome) of the molecules must also be considered. The triad then becomes a four-note harmony

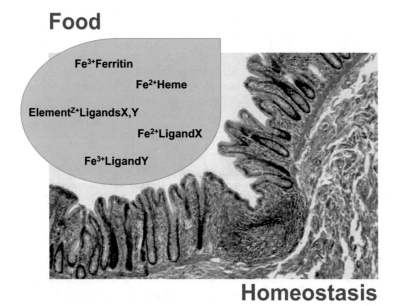

Fig. 7.2 Enterocyte with villi and the possible supply of different iron complexes in the diet: stable and labile Fe^{3+}- and Fe^{2+}-complexes with different inorganic and organic ligands, heme and ferritin iron, complexes of other bivalent metal ions ($z = 2$) such as zinc, which compete with Fe^{2+} at the DMT-1 transporter. (Right part of figure: © tonaquatic/stock.adobe.com)

receptor, zinc protoporphyrin and hepcidin are certainly also very important. A distinction must then also be made between parameters that determine long-term and short-term effects of bioavailable iron from food or from iron supplementation. In this context, it is also interesting to determine the haemoglobin content of the reticulocytes, the immediate pre-stage of the erythrocytes, which provides information on the temporal proximity of the body's iron supply.

In order to follow the path of iron into the organism more precisely, studies with the stable isotope ^{57}Fe are indicated, which then allow a calculation of the bioavailability under real conditions. Of course, the questions in which binding form the marker isotope ^{57}Fe is added to the food and which iron-containing molecules (haemoglobin, ferritin, transferrin, etc.), cells, cell compartments in the organism are selected for isotope determination by mass spectrometry must also be clarified. A very central question in all studies on bioavailability is: In which target organ, which target cell and which cell compartment or target molecule do I quantify my target element, i.e. in this case iron?

Equilibration experiments of the naturally occurring iron species mixture in a food with readily soluble ^{57}ferrous sulfate or ^{57}ferrous chloride could be attempted to enrich these original iron binding forms with ^{57}Fe, if the thermodynamics and kinetics of the various equilibrium reactions allow, which can then be verified by mass spectrometry. These prepared foods could then be used in human studies.

Furthermore, the use of defined isotope-labelled iron species that play an important role in foodstuffs, such as heme iron, ferritin iron, iron complexes with carboxylic acids or phenols, is interesting in this context. For example, a ferritin labeled with ^{57}Fe has been used for species-specific isotope dilution mass spectrometry (Hoppler et al. 2009). This standard, which is also of great interest for iron bioavailability studies, was produced by cloning and overexpressing the ferritin gene of Phaseolus vulgaris (common bean) in *E. coli* and culturing it in a medium containing ^{57}FeCl$_2$. This recombinant ^{57}Fe ferritin contains about 1000 iron ions per molecule with an isotopic enrichment of ^{57}Fe > 95% after isolation and purification (Hoppler et al. 2008).

Isotopes of Iron, Occurrence and Possible Uses
Stable isotopes and natural percentage composition of iron:
^{56}Fe: 91.72%, ^{54}Fe: 5.8%, ^{57}Fe: 2.2%, ^{58}Fe: 0.28%.

The radioactive isotope ^{59}Fe has been used in some studies on the absorption of iron (e.g. Theil et al. 2012; Lv et al. 2015). It is a β- and γ-emitter with a half-life of about 44 days and has important applications in medicine in ferrokinetics (iron clearance, iron utilization rate, erythrokinetics).

The stable isotope ^{57}Fe can be used as a universal standard in studies in iron biochemistry, detection is by mass spectrometry, the disadvantage of radioactivity in studies is eliminated (cf. Hoppler et al. 2008, 2009).

^{57}Fe has a nuclear spin of ½, which allows nuclear magnetic resonance spectroscopy (NMR) directly on the iron nucleus. This allows direct information to be obtained about the electronic states of iron atoms and ions in biologically relevant complexes. In NMR spectroscopy, the sample is placed

(*continued*)

(continued)
in a strong magnetic field, resulting in the formation of different energy levels
of the nuclear spin. By irradiation of electromagnetic waves of defined energy
these energy levels can be determined. The energy levels depend on the chemical environment of the measured nuclei and there are also couplings between
the nuclei. In ordinary NMR spectroscopy, the nuclei ^1H and ^{13}C are measured.

^{57}Fe also exhibits very good properties for Mößbauer spectroscopy, which
can be used, for example, to distinguish between iron (II) and iron (III) or to
make important statements about the chemical properties of ligands. In this
method, the resonance absorption of monochromatic gamma radiation by
atomic nuclei in solid samples is investigated. The high-resolution hyperfine
structure of the spectra obtained, which results from nuclear-electron interactions, can be used to make statements about the bonding states in the electron
shell. Rudolf Mößbauer was awarded the Nobel Prize in Physics in 1961 for
the discovery of the fundamentals of this investigation method.

For studies on bioavailability, the Caco-2 model (carcinoma colon) is very popular, which is also frequently used in the pharmaceutical industry to determine the
absorption of potential active ingredients. These colon cells form epithelial-like
monolayers that are tightly connected by dense bands of membrane proteins (tight
junctions), thus mimicking the barrier of the gastrointestinal tract. The monolayers
are located on a permeable membrane that separates the apical from the basolateral
chamber (Fig. 7.3).

The Caco-2 model is much more than a simple diffusion model. It has many
known carrier systems, some phase 1 and phase 2 enzymes and efflux pumps.
However, it is not small intestinal cells, is not perfused and does not have a mucus

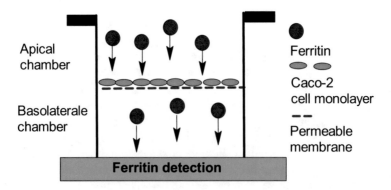

Fig. 7.3 Schematic representation of an experimental set-up of the Caco-2 model for the determination of the bioavailability of ferritin. The absorption of substances in the enterocyte can thus
be studied

layer. However, it has been extensively optimized and standardized since its introduction and is now a well-established in vitro test system used in many studies
(Ding et al. 2021).

7.1   Species Analysis of Iron in Foodstuffs

The binding situation of iron in the various plant and animal foods can be very complex, since in addition to two different oxidation states, a large number of possible
ligands are available for the iron ions. For some food groups, basic information on
the binding forms already exists, while for other products only theoretical ideas
based on complex chemical analogies exist but have not been experimentally
proven. Thus, there is still much to be done in the field of determination of the binding forms (species analysis) of iron in foods. A basic classification of iron species is
shown in Fig. 7.4.

Basically, the forms of iron binding are divided in the literature into heme iron
and non-heme iron, whereby the heme iron occurs mainly in animal foods. In most
works on bioavailability, no further differentiation is then made and the non-heme
iron is only considered summarily. However, instead, a further division of the non-
heme iron into ferritin iron and non-ferritin iron is useful and will be used below,
since ferritin iron is also a very stable and relatively defined iron species, for which
a separate transport system into the enterocytes also exists. The chemical properties
of ferritins are presented in Sect. 7.3.

In order to carry out a species analysis of elements in food, extracts must first be
prepared which contain the species as unchanged as possible and dissolved in large
proportions. This is called the initial step of species analysis. Only then are they
accessible to most sensitive and informative analytical-chemical techniques
(Günther 1997).

In addition, the uptake of the element of interest is also influenced by, among
other things, the presence of other elements and also their bonding forms. Therefore,
for a comprehensive nutritional assessment of an element in a given food matrix – in
addition to a determination of the binding form of the element of interest – information on absolute contents and binding states of other elements is also necessary.
These complex interrelationships lead to the necessity of multi-element speciation,
which is extremely important, especially in the case of foods (Fig. 7.5) (Günther 1997).

The first fundamental work on the initial step of species analysis and subsequent
multi-element species analysis was carried out in the 1990s on a wide range of
botanically very different food plants. In addition to iron, zinc, cadmium, copper,
calcium, strontium, manganese, potassium and rubidium were simultaneously analyzed and general binding principles of the elements were worked out (Günther 1997).

The initial step consists of cell disruption followed by ultracentrifugation and the
determination of the elemental fractions that are subsequently present in the super-
natant. This soluble fraction of the elements is now directly accessible to further
species analysis by coupling chromatographic and atomic spectrometric methods.
Figure 7.6 shows an example of the result for iron in a series of plant foods.

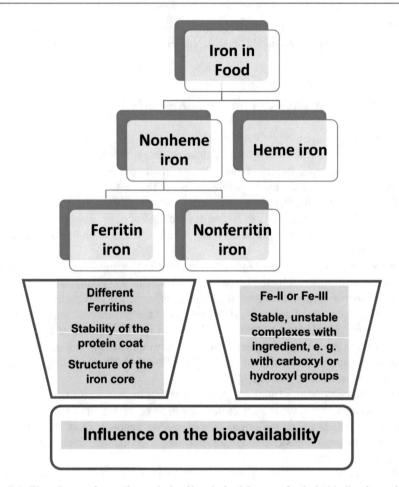

Fig. 7.4 Flow diagram for species analysis of iron in food. In many foods the binding form of iron is still unknown

The homogenization by the Ultra-Turrax® treatment simulates a chewing process and the destruction of the cell structure that takes place. This food-chemical approach of the investigations is very important, because possibly the iron ions, which are originally present as thermodynamically labile complexes, are bound differently after decompartmentalization in the case of a changed ligand supply. This new species composition then finds its way into the gastrointestinal tract. In contrast, for questions with a plant physiological background, one would first separate the cell organelles by ultracentrifugation and then perform an element species analysis, if possible preserving the originally present element binding form in the plant cells.

The percentages of iron in the individual cytosols of the foods ranged from 6% to 43% (Fig. 7.6). In the case of iceberg lettuce, spinach, chard, lettuce, red cabbage and especially kohlrabi, the cell disruption technique used was able to transfer 29–43% of the metal contained in the plants into solution. In radish, pepper, endive, dill, Jerusalem artichoke, cut lettuce and root parsley, iron was found to be less than

Fig. 7.5 Elements and element species influence each other during absorption in the organism. For this reason, multi-element speciation is necessary for the nutritional and toxicological evaluation of element contents in foods

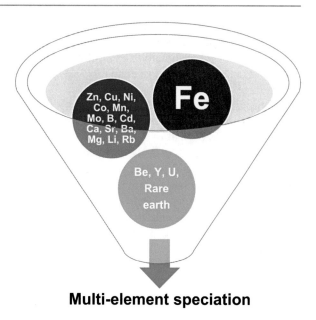

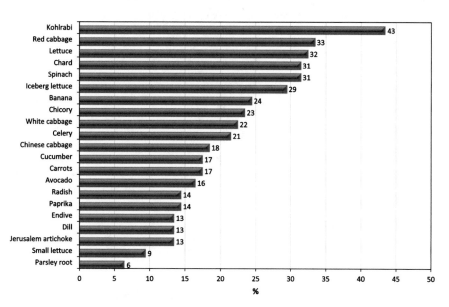

Fig. 7.6 Soluble fractions of iron in various plant foods in the supernatants (cytosols) after Ultra-Turrax® treatment (homogenization) in buffer and subsequent ultracentrifugation. First presentation of the concept of the initial step of elemental species analysis in foods (Günther 1997)

15% of the total content in the cytosol. The mean value of cytosolic proportion across all plant foods studied is only 21% for iron, which is much lower than, for example, zinc (68%), cadmium (51%), rubidium (92%), potassium (90%), copper (73%), manganese (35%) and calcium (26%). Only strontium, at 20%, has an even lower average cytosolic proportion than iron.

This means that, on average, iron in the foods investigated here is largely bound to insoluble pellet components, although there are of course exceptions, such as in kohlrabi, where almost half of the iron is present as soluble iron species (Fig. 7.6).

For further species analysis, the food extracts are then further separated and the fractions obtained are analyzed for elements using atomic spectrometry techniques. For the separation, stationary chromatographic phases should be used in particular, which show as little interaction with the analytes as possible in order to preserve the chemical structure of unstable iron species. As an alternative, field-flow fractionation can be used, which works without stationary phase. Furthermore, all operations should be carried out under protective gas such as nitrogen or argon to prevent the oxidation of Fe-(II) to Fe-(III) species by oxygen from the air.

For the detection of iron in the fractions after separation, the various techniques of atomic absorption spectrometry (AAS), such as flame AAS or graphite furnace AAS, can be used. Multi-methods such as total reflection X-ray fluorescence analysis (TXRF) or inductively coupled plasma mass spectrometry (ICP-MS) offer the possibility of simultaneously detecting many other elements as well (Fig. 7.7).

Fig. 7.7 Simplified representation of the iron species analysis in a food extract by a combination of gel permeation chromatography (separation according to molecule size) and ICP tripl. Quad. mass spectrometry (determination of iron and other elements, multi-element species analysis)

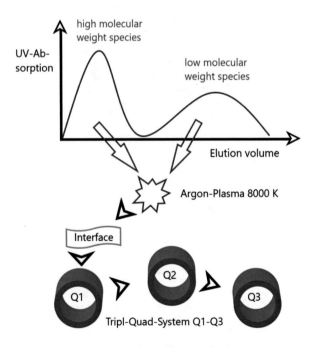

This enables multi-element species analysis in foodstuffs, which is the only way to gain a comprehensive insight into the elemental species system. With this knowledge, the mutual influence of the absorption of minerals and trace elements in the organism can then be better understood. Concepts to this effect were first presented in the 1990s (Günther 1997). Examples of results of an iron species analysis in various plant foods are shown in Fig. 7.8.

The fractions from the different molecular weight ranges of the gel permeation chromatography in which iron is present are then separated by further methods until the iron species are finally chemically pure. By methods of organic-chemical analytics, the substance class of the ligands of the iron complexes is first determined, such as e.g. heme proteins, ferritins, other Fe proteins, amino acids, polysaccharides, nucleic acids, polyphenols, carboxylic acids, in order to then determine the exact chemical structure by techniques of modern instrumental analytics (organic mass spectrometry, infrared spectroscopy, nuclear magnetic resonance spectroscopy).

Examples of frequently discussed ligands for iron complexes in foods include phytic acid, oxalic acid, citric acid or polyphenols (cf. Figure 7.9). These are low molecular weight ligands and could, for example, be components of iron species with a molecular weight less than 50 kDa (Fig. 7.8). For the high-molecular iron binding forms of 75–700 kDa, e.g. proteins, nucleic acids, polysaccharides or possibly ferritins could be considered, which are interesting research questions for the future (cf. Figure 7.8).

Due to the highly complex chemical composition of foodstuffs, there are certainly several hundred different compounds that are suitable for the complexation of Fe^{2+} and Fe^{3+} ions due to their functional groups. This is still a wide open field for future research activities, since the binding situation of iron and other elements in food is still largely unknown.

Fig. 7.8 Detected iron species in the cytosol of various foods by off-line coupling of gel permeation chromatography (Sephacryl-S-400, separation range 50–8000 kDa) with ICP mass spectrometry (Muktiono 2006)

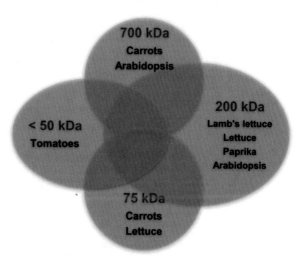

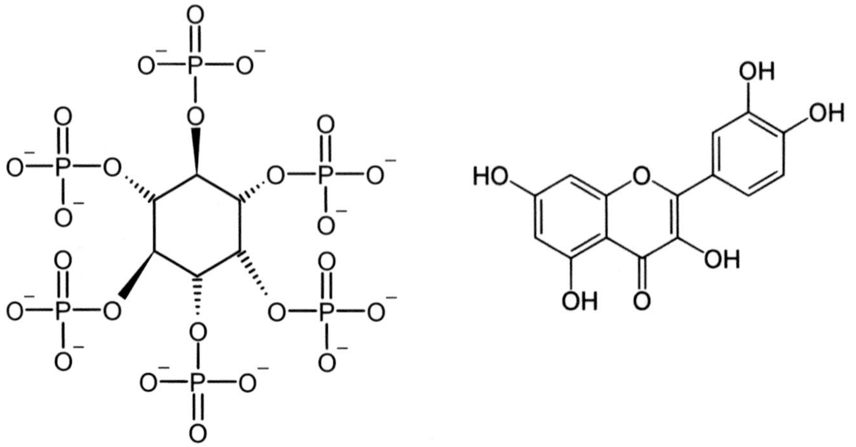

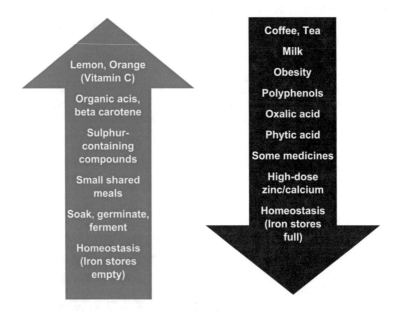

Fig. 7.9 Anion of phytic acid, an inositol phosphate (left) and quercetin (right), a known polyphenol from vegetables and fruit

Fig. 7.10 Selection of ways to increase (green) and decrease (red) the uptake of iron in the body

7.2   Dietary Components and Homeostasis

In addition to the original binding form of iron in food, the interactions of the element with other potential complexing agents present naturally also play a role and, in addition, homeostasis is an important factor in iron absorption (Fig. 7.10).

Vitamin C (cf. Fig. 3.6) converts iron-III, which cannot be absorbed by the body, into readily absorbable iron-II in a transmembrane process involving duodenal cytochrome b (cf. Sect. 3.2) and also binds it as a complex. These processes are responsible for a greatly increased absorption of iron. In cooked dishes, the vitamin should be added after the cooking process and stirred well. This ensures that the vitamin C is not destroyed by the heat. When making smoothies, on the other hand, it is better to add vitamin C, e.g. in the form of lemon juice, before mixing. This ensures better contact between the vitamin and the iron content of the plant cells. Lemons and oranges contain around 50 mg/100 g of edible vitamin C, whereas e.g. peppers, Brussels sprouts, kale and broccoli usually contain 100–150 mg/100 g of ascorbic acid. The intake recommendations of the German Nutrition Society for vitamin C for adults are 95 mg/day (w) and 110 mg/day (m).

Other organic acids such as malic acid, citric acid or lactic acid from fermented foods such as sauerkraut are also said to positively influence iron absorption. The situation is similar with beta-carotene, which is present in carrots at approx. 7 mg/100 g, but is also found in relevant quantities in kale, spinach, lamb's lettuce and chard.

The consumption of onions, leeks or garlic is also thought to increase iron absorption, for which the typical sulfur-containing compounds in these foods are held responsible. As an example, the allicin from garlic should be mentioned here, which is formed from the S-allyl-cysteine-sulfoxide (alliin) after destruction of the cell structure (i.e. during cutting or chewing) by enzymatic action and is supposed to cause an increase in iron absorption due to its chemical structure and also the sulfur-containing secondary products (garlic odor).

Furthermore, a distribution of meals throughout the day is very favorable for good iron absorption in the body, because the biochemical mechanisms of iron absorption are then more evenly utilized. Obesity-induced subclinical inflammation, on the other hand, is thought to reduce the absorption of iron.

Depending on the food, kitchen techniques such as sprouting, soaking or fermenting can substantially reduce the phytic acid content by activating the phytase enzyme, which is important in degradation. Phytic acid contains six phosphate groups via which iron ions can be complexed, which then impedes absorption (cf. Figure 7.9). For this purpose, pulses or whole grains can be soaked overnight before cooking and then allowed to germinate. The fermentation process, e.g. in the production of sourdough bread or fermented tofu, also ensures a breakdown of phytic acid and thus a higher bioavailability of the non-ferritin iron. Given the proportion of iron bound as ferritin in tofu (38% in soybeans, see Table 7.1), complexation with phytic acid should not play a role.

For a more in-depth discussion of the biochemically very complex topic of resorption-promoting and resorption-inhibiting aspects in iron, relevant review articles can be consulted (Zhang et al. 2021; Milman 2020; Collings et al. 2013).

The influence of the iron status and the hepcidin level in the organism on the percentage absorption of iron from food has already been discussed in more detail in Sects. 3.3 and 6.1. A high level of ferritin or hepcidin reduces the bioavailability of the element. A low level of both compounds increases the absorption of iron in the body. Inflammatory processes in the body increase the level of hepcidin and can thus also be responsible for iron deficiency (Prentice et al. 2017).

Table 7.1 Iron content and percentage of ferritin-bound iron in different legumes, ordered by descending percentage of ferritin-bound iron

Food	Iron content (mg/100g)	Ferritin iron (%)
Lenses	5.94	69
Yellow peas	4.52	62
Green peas	4.58	52
Soybeans	6.58	38
Pinto beans	5.56	29
Kidney beans	6.44	15

Total contents were determined by electrothermal atomic absorption spectrometry (ET-AAS), and analysis of the percentage of ferritin iron was performed by species-specific isotope dilution mass spectrometry, a recently developed and very precise method (Hoppler et al. 2009)

When using ferritin as a marker for iron status, it is important to consider any inflammatory processes that may be present in the organism that may influence the ferritin value. This can be done via inflammatory markers such as C-reactive protein (CRP). Taking these relationships into account, models have been developed that can be used to predict rates of iron absorption as a function of ferritin concentration (Fairweather-Tait et al. 2017). A comprehensive presentation on biomarkers for determining iron status including information on how the bioavailability of the element is influenced by different parameters and substances was published by Lynch et al. (2018).

Coffee and tea and some other plant foods contain complexing agents that can bind iron and thus reduce its absorption. Therefore, if possible, you should not drink coffee or tea 1 h before or after meals if you suffer from iron deficiency. The high calcium content of milk and dairy products can reduce the absorption of iron.

Polyphenols, oxalic acid or phytic acid should serve as examples here which can strongly complex iron cations via phenolic hydroxyl groups, carboxyl or phosphate groups and thus impede the resorption of the trace element.

Some drugs can also negatively influence the bioavailability of iron. These include, for example, bisphosphonates for the treatment of osteoporosis, antacids used against heartburn or neomycin, a broad-spectrum antibiotic (Gröber 2011).

A simultaneous high-dose supply of divalent cations such as Zn^{2+} or Ca^{2+} in the organism, as occurs, for example, after the ingestion of corresponding food supplements, can also hinder the uptake of iron as Fe^{2+} through competitive reactions at the divalent metal transporter (DMT-1, cf. Figs. 3.1 and 3.7). However, at natural and moderate physiological concentrations of other divalent cations, this effect should not play a role.

7.3   Properties and Uptake of Ferritin

Numerous recent studies have shown that there is a separate uptake pathway for ferritin iron that is independent of the uptake mechanisms for iron-II and heme iron (cf. Figure 7.11).

In many legumes, much of the iron is present as a ferritin complex (Masuda et al. 2017), and therefore a separate section is devoted to this particular iron species. In the currently most valid view in the scientific field, ferritin iron is thought to be absorbed directly into enterocytes. The discovery of this new, third uptake mechanism may represent a further departure from the conventional view that plant iron is inferior to iron from animal products for nutritional purposes. These new findings are the subject of intensive research at the moment. Figure 7.11 shows a schematic diagram of the three uptake pathways into the intestinal cell.

The plant ferritin iron remains intact when absorbed through the newly discovered ferritin port, similar to the readily absorbable heme iron that is introduced through the HCP-1. It is only in the intestinal cell that it is broken down and the iron is then available for its many essential functions. The uptake of iron through the ferritin port is thought to be slower than through the other two pathways and thus the process is more controllable by the cell and the iron supply is more consistent.

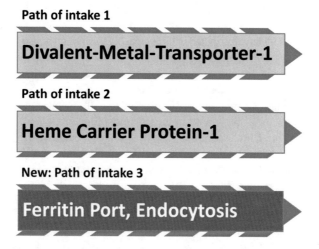

Path of intake 1

Divalent-Metal-Transporter-1

Path of intake 2

Heme Carrier Protein-1

New: Path of intake 3

Ferritin Port, Endocytosis

Fig. 7.11 The three uptake mechanisms for iron into enterocytes. Free iron-II enters the intestinal cells through DMT-1 (divalent metal transporter-1). Heme iron enters through HCP-1 (heme carrier protein-1). The resorption of ferritin iron occurs through the newly discovered ferritin port

The extent to which this process is subject to control over the iron status of the organism, as is generally the case with non-heme iron (cf. Chap. 6) needs to be investigated in the future. For this purpose, ferritin iron should be considered separately within the non-heme iron, as has already been suggested in the scheme for an iron species analysis (Fig. 7.4).

In ferritin absorption, a large number of iron ions are absorbed simultaneously per molecule; secondly, iron absorption is independent of the other two pathways; and thirdly, food components that make iron absorption difficult (cf. Figure 7.10) have no influence on the absorption of ferritin iron, since it is protected against external influences by the protein coat. Fourthly, this makes it clear that nature has certainly also provided its own absorption pathway for plant ferritin iron, which has played a role in human nutrition since early human history through the consumption of legumes.

At the beginning of the 2000s, the first summary publications by American research groups presented the current status and a positive assessment of the iron supply of the organism by ferritin (Theil 2004). Thereupon, the topic received very strong tailwind and was also intensively researched by other groups and the further very positive results were summarized in an article published in 2010 (Zhao 2010). The final confirmation that the uptake of ferritin iron is independent of the absorption of heme iron and iron-II (cf. Figure 7.11) was finally published in 2012 and gave further impetus to research in this area (Theil et al. 2012).

Further extensive studies showed a very good absorption property of plant ferritin in the organism (Lv et al. 2015; Khurana et al. 2017), and in recent years, widespread recommendations to use phyto-ferritin for the treatment of iron deficiency finally followed (Zielinska-Dawidziak 2015; Masuda et al. 2017), and corresponding US patent applications were filed (Theil 2017).

Ferritins consist of 24 subunits. They are bundles of four parallel and antiparallel alpha helices, the structure is quite similar in different species, and they are thus highly conserved biochemical structures (Fig. 7.12).

Ferritin was first isolated and crystallized from the spleen of the horse (Laufberger 1937). The protein is widely distributed in biological organisms and is also found in many plants including algae and in microorganisms. In legumes, a large proportion of the iron is often bound as ferritin. In animal organisms it is found mainly in the liver, spleen, bone marrow and blood. In humans, about 20% of the total iron is stored as ferritin.

The ferritin subunit has the shape of a cylinder with 5 nm length and 2.5 nm diameter. The diameter of the cavity in the complete ferritin molecule is 8 nm and the outer diameter is 12 nm, and it has various symmetry elements (Fig. 7.13).

The 24 subunits of the various ferritins differ in fine structure. Animal ferritin consists of the L-type (light, 19.5 kDa), which is important for the nucleation process of iron, and the H-type (heavy, 21.0 kDa), which has ferroxidase activity and is responsible for the oxidation of Fe-II to Fe-III. The ratio L/H in the whole molecule depends on the cell type.

Plant ferritin consists of 24 subunits of the H-1 type (26.5 kDa) and H-2 type (28.0 kDa), both of which possess ferroxidase activity. Both types have so-called EPs (extension peptides, cf. Figure 7.12) at the N-terminal end, which are about 30 amino acids in length and are localized on the surface. The EP have a protease

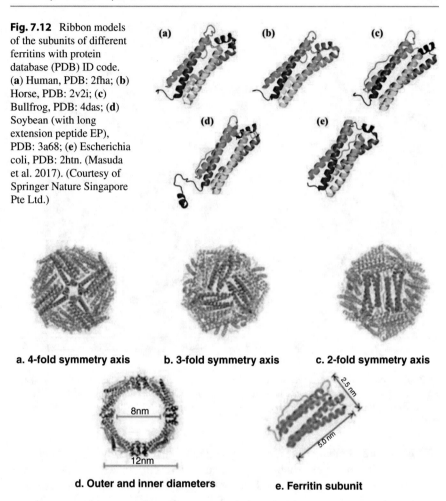

Fig. 7.12 Ribbon models of the subunits of different ferritins with protein database (PDB) ID code. (**a**) Human, PDB: 2fha; (**b**) Horse, PDB: 2v2i; (**c**) Bullfrog, PDB: 4das; (**d**) Soybean (with long extension peptide EP), PDB: 3a68; (**e**) Escherichia coli, PDB: 2htn. (Masuda et al. 2017). (Courtesy of Springer Nature Singapore Pte Ltd.)

Fig. 7.13 Ferritin protein coat showing the view of the different symmetry axes (**a–c**), with inner and outer diameters (**d**) and the representation of a cylindrical subunit with its length and diameter (**e**). (Masuda et al. 2017). (Courtesy of Springer Nature Singapore Pte Ltd.)

activity that destroys the ferritin coat during seed germination, thus releasing iron for metabolism (auto-degradation process), and when iron supply is high, the EP provide an additional mechanism of iron nucleation in ferritin. On the inner surface, ferritin has a strong negative charge, while the net charge on the outer surface is close to zero or slightly positive at pH 7 (Masuda et al. 2017).

The initial step of iron nucleation proceeds via a binuclear mechanism in which various specific binding sites are involved. Iron is complexed via the oxygens of the carboxyl groups of the glutamic acid side chains and via a nitrogen atom of the aromatic imidazole ring of histidine (Fig. 7.14).

As shown above, the shell of individual ferritins from different species has been relatively well studied, and the results show a high degree of similarity and many

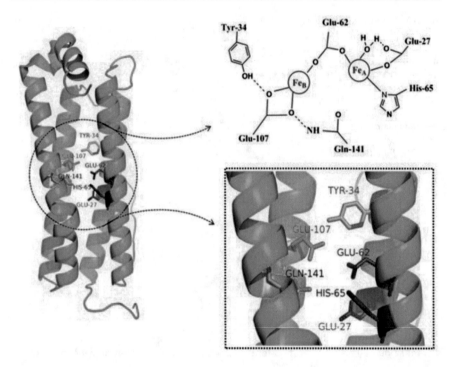

Fig. 7.14 Schematic view of the di-iron ferroxidase center of the four helices of H-(heavy)-ferritin. The ferroxidase center consists of iron binding sites A and B and the conserved amino acid ligands Glu27, His65; Glu62, Glu107, Gln141 and Tyr34. (Masuda et al. 2017). (Courtesy of Springer Nature Singapore Pte Ltd.)

common features, both in the quaternary structure and in the 24 subunits. In contrast, the iron core has been less studied to date, and the results suggest that this shows greater heterogeneity in structure and composition depending on the species studied.

More detailed information is available, for example, on the iron nuclei of ferritins from humans, bacteria, horses, rodents, molluscs or peas. The average number of iron ions per ferritin molecule varies between 800 and 2500 and the average diameter of the nucleus between 4.1 and 8.0 nm. The crystallinity of the nuclei is very good in some cases, but amorphous structures are also found. The molar ratio of Fe-(III) ions to intermediate phosphate (iP) is highly variable, and Fe/iP values between 21 and 1.5 are found. The exact structure of ferritin iron nuclei from different organisms is still a rather open field of research, where especially the use of modern methods of physical chemistry and spectroscopy is indicated in the future. Furthermore, the magnetic properties are also very interesting and will certainly be further focused on in the future.

The currently accepted conception of the structure of the crystalline iron core in ferritins can be approximated by the formula $5Fe_2O_3 \times 9H_2O$. Other studies using total X-ray scattering pair distribution function analysis methods led to the proposal of the molecular formula $Fe_{10}O_{14}(OH)_2$, where 80% of the iron ions are octahedrally surrounded by oxygen (Masuda et al. 2017).

The thermal stability of ferritin from different species is very good and an important property of this highly interesting protein quaternary structure of alpha-helices. Most studies report a denaturation temperature of >80 °C or even 90 °C, which is not affected by the presence or absence of the iron core. The properties of the EPs (extension peptides, cf. Figure 7.12) seem to influence the heat stability in this case. Even in heat-treated soy milk or tofu derived from it, the original structure of ferritin is largely intact, as confirmed by circular dichroism studies, and the cavity of the molecule still contains the iron core. This predestines tofu to be an excellent plant food for supplying the organism with ferritin iron (Masuda et al. 2017).

Circular dichroism (CD) measures the absorption of circularly polarized light. CD spectra allow a rapid characterization of the secondary structure of proteins in particular, i.e. in the case of ferritin the alpha helices.

Overview

Production of tofu:

The proteins in soybean milk are precipitated by adding mineral salts and then filtered off and pressed (soybean curd). Soy milk is made by soaking and pureeing yellow soybeans. After filtration of the mass through a fine sieve or cloth, soy milk is obtained, which must be boiled for another 15 min on a low heat to eliminate various antinutritive factors. These include in particular digestion-inhibiting trypsin inhibitors and lectins (hemagglutinins). To make tofu, either magnesium chloride ($MgCl_2$, Nigari, Japanese), calcium sulfate ($CaSO_4$, Chinese), or a mixture of both salts is added to the soy milk, resulting in precipitation of the proteins. When the bean curd is pressed through a fine-mesh cloth, the tofu is then obtained. The finished product contains about 5 mg iron/100 g.

In addition to its resistance to thermal treatment, the chemical structure of the ferritin molecule is also stable to other preservation methods used in food technology, such as HHP technology (high hydrostatic pressure, high-pressure pasteurization). In this process, the products and components are subjected to a hydrostatic pressure of up to 6000 bar without the use of heat or other additives. This eliminates germs and bacteria and avoids the disadvantages of heat treatment, such as the loss of vitamins or aromatic substances.

The determination of the amount of ferritin-bound iron in food is not trivial. Mostly, methods based on the principle of immunoassay have been used for this purpose, differing in various end detection techniques (turbidimetry, etc.). Of course, the ferritin protein coat is targeted as the analyte, and no statement can be made about the iron content in the cavity, which sometimes varies considerably. By contrast, species-specific isotope dilution mass spectrometry using [57]Fe-ferritin as a standard can be used to accurately determine the exact ferritin iron content in food-stuffs (Table 7.1).

Among the legumes studied, lentils take the top spot. Almost 70% of the iron contained in them, at approx. 6 mg iron/100 g total content, is present as ferritin iron, which is then 4.5 mg iron in the cavity with its own enterocyte port per 100 g lentils. The binding form of the remaining iron is as yet unknown. However, uptake should also be possible here after reduction to iron-(II) by vitamin C (cf. Figure 7.15).

The uptake of ferritin occurs by clathrin-mediated endocytosis, is independent of the uptake of ferrous salts and heme iron, and shows good bioavailability (Theil et al. 2012; Lv et al. 2015; Masuda et al. 2017; Khurana et al. 2017; Theil 2017; Perfecto et al. 2018). Therefore, plant ferritin is also recommended as an excellent preparation for the treatment of iron deficiency states due to its good absorption in the organism and its special tolerability (Zielinska-Dawidziak 2015).

Through the consumption of ferritin-rich plants, such as legumes, many antioxidant secondary plant substances are also absorbed. They reduce oxidative stress in the organism and thus counteract inflammatory processes in the body. This then leads to a drop in hepcidin levels, and iron absorption increases (Prentice et al. 2017) (cf. Figs. 3.10 and 3.11).

Phyto-ferritin therefore has several advantages and will certainly play an even greater role in the supply of iron in the future with the population's shift towards a more plant-based diet, which is indicated in many respects. In this context, studies with food-technologically interesting protein compositions using phyto-ferritin for the manufacture of products for the future market of vegetarian and vegan foods are certainly also very interesting.

Clathrin is a coat protein involved in cell membrane invagination and vesicle formation (Fig. 7.16). Clathrin-mediated endocytosis was discovered over 50 years ago and is a key process in the vesicular transport of a wide range of cargo molecules from the cell surface to the cell interior. Over 50 proteins are part of the machinery that generates the endocytic clathrin-coated vesicles (Kaksonen and Roux 2018).

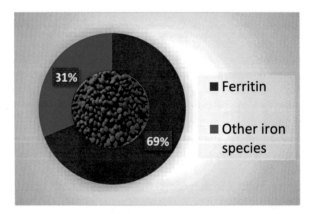

Fig. 7.15 Iron binding forms in lentils. 69% of the iron contained is bound as ferritin. The remaining part is probably present as Fe-(III) complex with low molecular weight ligands. This iron fraction can be taken up by the DMT-1 after reduction by ascorbic acid with the participation of DCytb. However, for a precise chemical structure analysis of the remaining 31%, further iron species analysis will be necessary in the future (cf. Sect. 7.1)

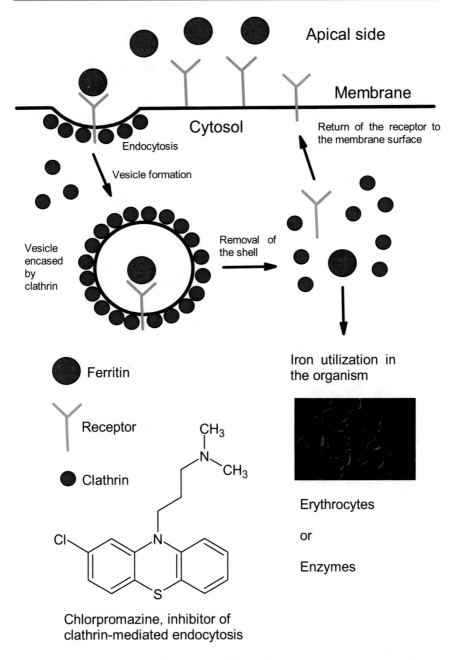

Fig. 7.16 Proposed mechanism for the uptake of ferritin by receptor clathrin-mediated endocytosis. In the presence of chlorpromazine, a known inhibitor of this type of endocytosis, absorption of ferritin is impeded in the Caco-2 model (see Fig. 7.3). (Perfecto et al. 2018). (Erythrocytes: Copyright © flashmovie/stock.adobe.com)

The interesting properties of the ferritin molecule are now used in many different disciplines. For example, the cage of ferritin becomes permeable in the presence of urea, and the iron core can be replaced by other substances such as drugs or other trace elements. The cage then protects the substances inside from attack of whatever kind, making it a very interesting transport molecule for sensitive nutritive components (Yang et al. 2015).

Epigallocathechin gallate, a medically interesting green tea polyphenol and antioxidant, could be moved into the cavity of apoferritin in this way (Yang et al. 2017), and there are now countless potential applications throughout nanotechnology, nanomedicine, and nanopharmacy that take advantage of the special properties of the ferritin molecule (Jutz et al. 2015).

References

Collings R, Harvey L, Hooper L, Hurst R, Brown T, Ansett J, King M, Fairweather-Tait J (2013) The absorption of iron from whole diets: a systematic review. Am J Clin Nutr 98:65–81

Ding X, Hu X, Chen Y, Xie J, Ying M, Wang Y, Yu Q (2021) Differentiated caco-2 cell models in food-intestine interaction study: current applications and future trends. Trends Food Sci Technol 107:455–465. https://doi.org/10.1016/j.tifs.2020.11.015

Fairweather-Tait S, Jennings A, Harvey L, Berry R, Walton J, Dainty J (2017) Modelling tool for calculation dietary iron bioavailability in iron-sufficient adults. Am J Clin Nutr 105:1408–1414

Gröber U (2011) Mikronährstoffe: metabolic tuning – Prävention – Therapie, 3. Aufl. Wissenschaftliche Verlagsgesellschaft, Stuttgart

Günther K (1997) Beiträge zur Multielement-Speziation in pflanzlichen Lebensmitteln: Studien zur Bindungsform zahlreicher Elemente unter besonderer Berücksichtigung von Zink und Cadmium, Habilitationsschrift Universität Bonn, Berichte des Forschungszentrums Jülich 3358, ISSN 0944-2952

Hoppler M, Meile L, Walczyk T (2008) Biosynthesis, isolation and characterization of ^{57}Fe-enriched Phaseolus vulgaris ferritin after heterologous expression in Escherichia coli. Anal Bioanal Chem 390:53–59

Hoppler M, Zeder C, Walczyk T (2009) Quantification of ferritin-bound iron in plant samples by isotope tagging and species-specific isotope dilution mass spectrometry. Anal Chem 81:7368–7372

Jutz G, van Rijn P, Miranda B, Böker A (2015) Ferritin: a versatile building block for bionanotechnology. Chem Rev 115:1653–1701

Kaksonen M, Roux A (2018) Mechanisms of clathrin-mediated endocytosis. Nat Rev Mol Cell Biol 19:313–326

Khurana M, Fung E, Vichinsky E, Theil E (2017) Dietary nonheme iron is equally bioavailable from ferritin or ferrous sulfat in thalassemia intermedia. Pediatr Hematol Oncol 34:455–467

Laufberger V (1937) Sur la cristallisation de la ferritine. Bull Soc Chim Biol 19:1575–1582

Lv C, Zhao G, Lönnerdal B (2015) Bioavailability of iron from plant and animal ferritins. J Nutr Biochem 26:532–540

Lynch S, Pfeiffer C, Georgieff M, Brittenham G, Fairweather-Tait S, Hurrell R, McArdle H, Raiten D (2018) Biomarkers of nutrition for development (BOND) – iron review. J Nutr 148:1001S–1067S

Masuda T, Chen H, Zhao G (2017) Structure, function, and nutrition of ferritin from foodstuffs. In: Zhao G (ed) Mineral containing proteins: roles in nutrition. Springer, Singapore

Milman N (2020) A review of nutrients and compounds, which promote or inhibit intestinal iron absorption: making a platform for dietary measures that can reduce iron uptake in patients with genetic haemochromatosis. J Nutr Metab. https://doi.org/10.1155/2020/7373498

Muktiono B (2006) Beiträge zur Multielement-Speziation in pflanzlichen Lebensmitteln durch off-line-Kopplung von Gelpermeationschromatographie und Massenspektrometrie mit induktiv gekoppeltem Plasma. Dissertation Universität Bonn

Perfecto A, Rodriguez-Ramiro I, Rodriguez-Celma J, Sharp P, Balk J, Fairweather-Tait S (2018) Pea ferritin stability under gastric pH conditions determines the mechanism of iron uptake in Caco-2 cells. J Nutr 148:1229–1235

Prentice A, Mendoza Y, Pereira D, Cerami C, Wegmuller R, Constable A, Spieldenner J (2017) Dietary strategies for improving iron status: balancing safety and efficacy. Nutr Rev 75:49–60

Theil E (2004) Iron, ferritin, and nutrition. Annu Rev Nutr 24:327–343

Theil E (2017) Methods for isolation, use and analysis of ferritin, United States patent application publication No.: US2017/0087209A1

Theil E, Chen H, Miranda C, Janser H, Elsenhans B, Nunez M, Pizarro F, Schümann K (2012) Absorption of iron from ferritin is independent of heme iron and ferrous salts in woman and rat intestinal segments. J Nutr 142:478–483

Yang R, Zhou Z, Sun G, Gao Y, Xu J (2015) Ferritin a novel vehicle for iron supplementation and food nutritional factors encapsulation. Trends Food Sci Technol 44:189–200

Yang R, Liu Y, Meng D, Chen Z, Blanchard C, Zhou Z (2017) Urea-driven epigallocatechin gallate (EGCG) permeation into the ferritin cage, an innovative method for fabrication of protein-polyphenol co-assemblies. J Agric Food Chem 65:1410–1419

Zhang Y, Stockmann R, Ng K, Ajlouni S (2021) Opportunities for plant-derived enhancers for iron, zinc, and calcium bioavailability: a review. Compr Rev Food Sci Food Saf 20:652–685

Zhao G (2010) Phytoferritin and its implications for human health and nutrition. Biochem Biophys Acta 1800:815–823

Zielinska-Dawidziak M (2015) Plant ferritin – a source of iron to prevent its deficiency. Nutrients 7:1184–1201

Diet for Iron Deficiency

8

8.1 Iron Content in Foodstuffs

After an extensive review of the available, valid information on iron contents in foods and their assessment, extracts of it have been compiled in this chapter in the overview presentations for plant and animal foods. In addition, there is a list of products that often contain very little iron (<0.5 mg iron/100 g). Care has been taken to also add information on whether the food is fresh, dried or already processed. This is very important when estimating the iron intake from the food and is not listed in many publications on iron contents.

The internet portals of the United States Department of Agriculture (USDA) and the Federal Food Code (BLS) provide quick and comprehensive information on the iron content of other foods and products (USDA 2021; BLS 2021). Access to the database from USDA is free, while there is a fee to use the BLS. With Food Data Central, the USDA is pursuing an interesting approach to food ingredient information processing that links databases of different types. Eurofir, an international, membership-based non-profit association under Belgian law (Eurofir 2021), is being established to provide valid food information in Europe. Own investigations on contents of minerals and trace elements including iron have been performed over the years by total reflection X-ray fluorescence analysis (TXRF) (Klockenkämper and von Bohlen 2015), inductively coupled plasma mass spectrometry (ICP-MS) (Becker 2007) and atomic absorption spectrometry (AAS) (Welz and Sperling 1999).

In the overviews listed here, ranges are given because the contents in the products can naturally fluctuate more or less due to different natural conditions. Reminder: 14 mg iron is the reference amount for the daily intake of an adult (nutrient reference value).

Contents of iron in groups of various plant foods in mg/100 g, ordered by descending iron content (USDA 2021, own research).

- Thyme, basil, spearmint, marjoram (dried): 80–120
- Parsley, dill (dried): 46–54
- Cocoa powder, sugar-free (commercial goods): 20–40
- Spirulina algae (dried): 28–32
- Rhenish sugar beet syrup (commercial): 15–35
- Wheat bran (commercial): 15–17
- Pumpkin seeds (commercial): 10–14
- Sesame seeds, poppy seeds, pine nuts (dried): 9–13
- Chia seeds, millet, millet flakes, sesame seeds, tahina (commercial goods): 8–10
- Dark chocolate (commercial): 6–10
- Flaxseed (dried): 7–9
- various lentils, beans, oatmeal (dried): 5–10
- Soy flour, soy granules (commercial goods): 8–10
- Amaranth, quinoa, chickpeas (dried): 7–9
- Kidney beans (commercial): 6–8
- Cashew nuts (commercial): 5–7
- Chanterelles (fresh): 5–7
- Sunflower seeds, oat flakes (commercial): 4–6
- Peas (dried): 4–6
- Dill (fresh): 4–5
- Wholemeal flour (wheat, rye, commercial): 3–5
- Spinach (raw): 3–5
- Lentils (ready to eat), parsley (fresh): 3–5
- Walnuts, hazelnuts, Brazil nuts, almonds (commercial): 3–5
- Wholemeal bread, buckwheat (commercial): 3–4
- Chickpeas (raw, fresh), white beans (cooked): 2–3
- Peas (prepared), brown rice (commercial, wholemeal): 1–3
- Kale, root parsley, leek, garlic, lamb's lettuce, carrots, corn (raw, fresh): 1–2
- Savoy cabbage, beetroot, peppers (fresh): 0.4–1
- Currants, raspberries, strawberries (fresh): 0.8–1
- Tomatoes, cucumbers, cauliflower, kohlrabi, potatoes (raw): 0.3–0.7
- Bananas, apples, oranges, lemons, grapes, pineapple, kiwi, mango, onions (fresh, commercial): 0.1–0.5

The differences in iron content can be very large, in extreme cases by a factor of 1000 (e.g. apples, oranges and dried spices), although it must of course be borne in mind here that these are dried spices and are not normally consumed in large quantities, so do not contribute much to the iron supply when prepared in the usual way.

However, a difference in iron content by a factor of 10 is very often found, even in realistic comparisons: ready-to-eat lentils, for example, contain more than ten times as much iron as bananas or apples. This is enormous and shows very clearly that when looking at the causes of the development of iron deficiency, one must look in particular at the foods that are consumed in large quantities every day, but contain very little iron.

When looking at the composition of foods in the iron-negative list (cf. Sect. 8.2), it can also be seen that a difference in the iron content in a usual food basket (composition of foods consumed frequently on average) may well amount to a factor of 10,000 (!).

The highest iron contents are found in spices. The concentrations are considerable. 2 g thyme or 4 g parsley (Fig. 8.1) or dill – of course in dry form – thus provide approx. 14% of the daily required amount of iron. The 100% would already be reached with approx. 50 g spirulina, a microalgae and also trend food, which is nowadays on the menu in some households.

Fig. 8.1 Parsley dried: 50 mg iron/100 g (average). Very fluffy, a filled dinner plate weighs only approx. 5 g. (The comparison with Fig. 8.2 clearly shows the difference in iron content by a factor of approx. 10 (dry/fresh difference))

Fig. 8.2 Flat-leaf parsley
fresh: 4 mg iron/100 g
(average)

In the case of plant foods, in addition to spices and wheat bran (Fig. 8.3) with an impressive 15–17 mg iron/100 g, it is above all pulses that have a relatively high iron content and can also be consumed in larger quantities. Lentils, beans, peas or soya almost all contain 5–10 mg iron/100 g in relation to the dry product available for purchase (Figs. 8.4, 8.5, 8.6, 8.7, 8.8, 8.9 and 8.10).

Fig. 8.3 Wheat bran:
16 mg iron/100 g (average)

Fig. 8.4 Pardina lentils:
8 mg iron/100 g (package
information)

Fig. 8.5 Yellow lentils: 5.8 mg iron/100 g (package specification)

Fig. 8.6 Red lentils: 5.5 mg iron/100 g (package specification)

Fig. 8.7 Beluga lentils: 7 mg iron/100 g (package specification)

Fig. 8.8 White beans: 7.5 mg iron/100 g (average)

Fig. 8.9 Green peas with the shell broken open, where most of the minerals are concentrated: 5 mg iron/100 g (average)

Fig. 8.10 Soy granules for the simple preparation of tasty dishes such as vegetarian dumplings (cf. Figure 8.11) or fritters: 9 mg iron/100 g (average)

Fig. 8.11 Soy dumplings, raw. The preparation is very simple: let the granules soak for 20 min, add other ingredients and form dumplings

But that is not all. The members of this plant family contain many other important minerals, trace elements, vitamins, fiber and valuable protein. They also ensure a slow rise in blood sugar levels and are therefore particularly suitable for diabetics.

Bran (Fig. 8.3) is the term used to describe the components left over after sifting the flour during grain processing. These consist of husk portions, the aleurone layer (separates the endosperm from the outer husk) and the germ and thus contain many valuable minerals and trace elements in high concentrations.

Interesting Facts About Lentils

The edible lentils (Lens culinaris) are descended from the wild lentil (Lens orientalis) and are one of the oldest cultivated and useful plants in the world with very valuable ingredients. Their original home is Asia Minor, and today many different varieties are found all over the world. The oldest findings of lentils in food preparations date back to the Stone Age, and lentils are also mentioned as a food in the Old Testament. In ancient Egypt and ancient Rome, lentils were considered an important staple food.

The world's largest producer of lentils is Canada at the rate of 4 million ton per year, followed by India (1.3 million), Turkey (0.45 million) and the USA (0.35). Other growing areas are in Australia, Russia, China and Nepal. Special varieties come from Spain (Pardina lentils), France, Italy and Greece. Lentils are also cultivated in Germany in the Swabian Alps, where spaetzle with lentils are also eaten and are popular.

Lentils are excellent sources of nutrients. They contain up to 30% proteins with a high biological value. Other high proportions are carbohydrates with 45% and the beneficial dietary fiber with 15%. Due to the special binding forms of the carbohydrates (complex carbohydrates), they are digested more slowly, thus ensuring a slow rise in the blood sugar level and keeping it stable for a longer period of time. That is why lentils have a low glycemic index.

Lentils contain the vitamins B_1, B_2, B_6, A and folate and are rich in lysine and the minerals magnesium, potassium, zinc and manganese. They contain up to 8 mg iron/100 g dry weight, more than two-thirds of which is bound as ferritin (cf. Fig. 7.15), making them an excellent source of iron.

Lentils come in a very wide variety and in different sizes between 4 and 8 mm and the main colors brown, red, yellow and green. In addition, there are black Beluga lentils from North America, which have a chestnut aroma, and special varieties such as the green-black speckled Puy lentils, which are grown on volcanic soils in France and taste very aromatic. Red and yellow lentils are not special varieties, but are usually the shelled form of brown lentils. As they no longer have a skin, they contain fewer minerals and therefore less iron, but cook very quickly and can be easily pureed.

The very well-known Pardina lentils have been grown in the Spanish north for centuries, retain their structure when cooked and are versatile in the kitchen. Mountain lentils is a collective term for lentils grown in mountainous regions around the world, and the light green-beige Alb lentils come from the Swabian Alb. Plate lentils are green to yellow-brown, quite large and are considered all-purpose lentils for rustic cooking. Smaller are the reddish-brown Chateau lentils, a French speciality from the Champagne region.

The Mexican chia (Fig. 8.12) was used as food by the Aztecs and Mayas as long as 5000 years ago, and is now often touted in the trade as a so-called "superfood." This has caused a great deal of controversy, and in summary this description can be described as far exaggerated. Certainly, chia seeds contain valuable ingredients, but significant differences to conventional, cheaper plant foods from regional cultivation have not yet been found.

The iron content of the pseudo cereal is with an average of 7–9 mg/100 g in the range of lentils and thus quite high. The neutral tasting chia seeds can be added to mueslis or smoothies, or also processed in pasta such as bread or pretzels.

Amaranth (Fig. 8.13) is also a pseudocereal and is often imported from Central or South America. The slightly nutty taste goes well with many savoury dishes and amaranth is suitable, for example, as a rice alternative. The iron content of 8–9 mg/100 g is comparatively high and is of the same order as lentils and chia seeds.

Quinoa (Fig. 8.14) also comes from South America and served as a staple food for the local population thousands of years ago. It is also a pseudo-cereal and is mainly grown where real cereals cannot thrive, such as in the Andes. This is one of the reasons why 2013 was declared the "Year of Quinoa" by the United Nations, as the plant is very robust and requires little water. In addition to many valuable ingredients, its iron content of 6–8 mg/100 g is also very remarkable. Quinoa is well suited as a side dish to fish or meat and is a tasty alternative to potatoes or pasta.

Fig. 8.12 Chia seeds: 8 mg iron/100 g (average)

Fig. 8.13 Amaranth: 8 mg iron/100 g (average)

Fig. 8.14 Quinoa: robust and very sustainable, 7 mg iron/100 g (average)

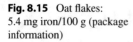

Fig. 8.15 Oat flakes:
5.4 mg iron/100 g (package
information)

Oat flakes are well known in European cuisine (Fig. 8.15). They are often made from the whole grain of seed oats and have a high content of valuable ingredients. They are commercially available as seeded flakes, tender flakes, or soluble melt flakes. The latter dissolve when stirred into liquid and are also suitable as a gentle food. As a wholemeal product, oat flakes have a relatively high iron content of 5–8 mg/100 g. Similar values are found in millet flakes (Fig. 8.16). Cereal flakes can be produced by the consumer with a commercially available flake crusher.

Sunflower seeds, pumpkin seeds, cashew nuts, walnuts, linseeds, hazelnuts or almonds generally contain a lot of iron, other minerals and also other value-giving ingredients. They can be used in a variety of ways, especially in tasty muesli mixtures, and thus contribute to the iron supply (Figs. 8.17 and 8.18). Linseed stimulates digestion through mucilage and contains small quantities of cyanogenic glycosides, from which hydrocyanic acid may be formed. Therefore, no more than 15 g per meal should be consumed.

Sesame is one of the oldest cultivated plants in the world. The seeds are used unhulled, hulled, unroasted or roasted in the kitchen. They consist of approx. 50% sesame oil with a relatively high proportion of unsaturated fatty acids, 20% proteins, 20% carbohydrates and 5% minerals. Sesame paste is popular for making hummus (Fig. 8.19).

Pine nuts are the shelled seeds from the cones of the pine tree (Fig. 8.20). They are very popular in Mediterranean cuisine for baking, in salads or in muesli. They

Fig. 8.16 Millet flakes:
9 mg iron/100 g (average)

are obtained either from the European pine (Mediterranean pine) or from the Korean pine from Asia.

Chocolate consists of cocoa mass, additional cocoa butter, milk and sugar. The cocoa mass is obtained from the cocoa beans and only it contains iron. Therefore, especially dark chocolate with a high cocoa mass content >70% is a very good source of iron, which has the further advantage that it contains little sugar (Fig. 8.21). Some varieties on the market contain 80%, some even 99% cocoa mass. As with all considerations of iron absorption in the organism, the bioavailability of the element plays a major role, which is strongly dependent on the type of cocoa product.

For sweet dishes and also cakes, it is advisable, if it is possible in terms of taste, to replace white household sugar with largely natural and regionally produced sugar beet syrup (Fig. 8.22). This makes it quite easy to enrich a wide variety of dishes with iron and other minerals such as potassium and zinc. Furthermore, the iron components in sugar beet syrup should be readily bioavailable due to its other constituents, as initial assessments have shown.

For a high iron intake, foods are of great advantage that can be consumed in large quantities, that have a high content of the element, and still bring many other value-giving ingredients at the same time. In addition, it is advantageous if they can be purchased relatively cheaply. This is particularly the case with lentils and beans, which are therefore a very good staple food in terms of iron supply. Lentils and beans come in many varieties, and very tasty dishes can be prepared with them. Therefore, one can assign a central position to these legumes in terms of iron supply.

Among animal foods, it is above all offal that is very rich in iron. In contrast, muscle meat from beef and pork does not contain as much iron as is often assumed. Poultry meat contains less iron on average and fish is often a poor source of iron, but it has other advantages in the diet, such as its high content of essential fatty acids. Milk and dairy products contain little iron (Figs. 8.23, 8.24 and 8.25).

Fig. 8.17 Sunflower seeds (5 mg iron/100 g), pumpkin seeds (12 mg iron/100 g), cashew nuts (6 mg iron/100 g), walnuts (4 mg iron/100 g), linseeds (8 mg iron/100 g), hazelnuts (4 mg iron/100 g). (All average values)

Fig. 8.18 Chopped
almonds: 4 mg iron/100 g
(average)

Fig. 8.19 Tahini sesame
paste: 9 mg iron/100 g. It
can be prepared by
mechanically crushing
hulled sesame seeds in a
high-powered blender with
the addition of a little
sesame oil

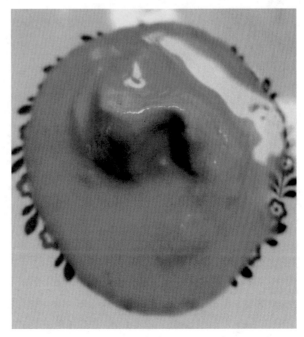

Fig. 8.20 Pine nuts: 11 mg iron/100 g (average). They have a very soft consistency and can serve as a convenient source of iron in many dishes

Fig. 8.21 Left: dark chocolate: 8 mg iron/100 g (average). Right: chocolate for garnishing and baking. Dark chocolate in particular is a good source of iron, as it contains a high proportion of cocoa mass

Fig. 8.22   Left: Sugar beet syrup: 25 mg iron/100 g (average). No relevant quantities of trace elements are found in refined sugar. Right: The ratio of the concentrations of potassium/sodium in sugar beet syrup is approx. 10, which is nutritionally very advantageous. This is a major advantage over molasses, where the levels of sodium are much higher than those of potassium

Fig. 8.23   Beef, steak fresh: 2.5 mg iron/100 g (average)

Fig. 8.24 Poultry: 1.5 mg
iron/100 g (average)

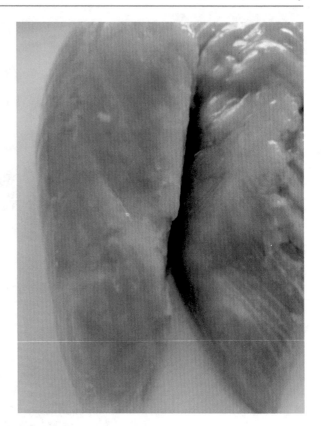

Fig. 8.25 Salmon: 1 mg iron/100 g (average)

Levels of iron in groups of various animal foods including dairy products in mg/100 g in commercial products. Ordered by descending iron content (USDA 2021)
- Pork liver: 15–30
- Calf's liver, beef liver: 7–8
- Blood sausage, liver sausage: 5–7
- Corned beef: 4–5
- Beef: 2–3
- Poultry, chicken leg: 1–3
- Pork, veal (cutlets): 2–3
- Oysters: 5–6
- Mussels: 4–6
- Egg yolk, chicken: 3–4
- Whole egg, chicken: 1.5–2.5
- Tuna: 1–2
- Bacon (pork, raw, smoked): 0.5–1.5
- Salmon, cod, herring: 0.2–2.0
- Cheese, egg white chicken egg: 0.1–0.3
- Milk: 0.05–0.15

8.2   The Iron Negative List

When studying the extensive scientific literature on the iron content in foods, it is noticeable that the ranges of variation for some products (e.g. fish) are very large and many foods classified as "very healthy" contain very little iron (Figs. 8.26 and 8.27). A selection of these foods is shown in the iron negative list.

If these products in particular are given special preference in the composition of a nutrition plan, individuals or patients can quickly fall into an undersupply of iron. This can be an important point in nutritional counselling that is often not taken into account, as the focus nowadays is often on other nutritively positively effective components such as vitamins or antioxidants.

Fig. 8.26 Vegetables and fruit often contain less than 0.5 mg iron/100 g fresh produce. This can lead to hidden iron starvation, with the numerous symptoms associated with it

Fig. 8.27 Mixed vegetable salat often contains very little iron, but of course it contains many other valuable ingredients such as vitamins or fiber. A large portion of 300 g usually contains less than 1 mg of iron (less than 10% of the reference amount)

Iron Negative List: Examples of foods with 0.5 mg of iron or less in 100 g of commercial product. These low levels are quite common in products. Ordered by descending iron content (USDA 2021, own research). This list is only a small selection of the possible products with very low iron contents.

- Artichokes: 0.5
- Rusk: 0.5
- Noodles (normal): 0.5
- Rice (brown): 0.5
- Tomatoes (raw): 0.5
- Mushrooms (fresh): 0.5
- Avocado (raw): 0.49
- Peppers (fresh): 0.48
- Leeks (fresh): 0.45
- Honey: 0.42
- Chicory (fresh): 0.42

(continued)

(continued)
- Jerusalem artichoke (raw): 0.41
- Kohlrabi (raw): 0.4
- Wine: 0.37
- Cherries: 0.36
- Radish (raw): 0.34
- Iceberg lettuce (fresh): 0.34
- Soy sauce: 0.33
- Cauliflower (cooked): 0.32
- Red cabbage (raw): 0.32
- Celery (raw): 0.32
- kiwi fruit: 0.3
- Chicken breast: 0.3
- Carrots (raw): 0.3
- Various cabbages: 0.3
- Bananas (raw, ready-to-eat): 0.26
- Grapefruit (raw, ready to eat): 0.26
- Guavas: 0.26
- Bananas: 0.25
- Potatoes: 0.25
- Papayas (raw): 0.25
- Peach (raw): 0.25
- Pineapple (raw): 0.25
- Milk chocolate: 0.25
- Pears (raw): 0.24
- White cabbage (raw): 0.24
- Vegetable cucumbers: 0.22
- Coffee drink: 0.2
- Mangoes (raw): 0.16
- Fish (Atlantic/Pacific, fresh): 0.16
- Tomato juice: 0.15
- Lettuce (fresh): 0.15
- Fruit smoothies: 0.15
- Various jams: 0.15
- White rice (cooked): 0.14
- Oranges (edible part): 0.1
- Apples (fresh): 0.1
- Pears (fresh): 0.1
- Household sugar (white): 0.06
- Ice cream: 0.06
- Pudding, ready to eat: 0.05
- Milk, buttermilk: 0.06

(continued)

(continued)

- butter: 0.05
- Yogurt: 0.05
- protein from hen's eggs: 0.04
- Orange juice and other fruit juices: 0.04
- Spirits: 0.04
- Beef and chicken broth: 0.02
- Cola drinks: 0.02
- Energy drinks: 0.02
- Sweets: 0.02
- Kitchen oils: 0.01
- Tonic water: 0.01
- Tea drink: 0.01
- Various types of cheese: <0.01
- Ready-made baby food from various manufacturers: < 0.01
- Beer: < 0.01
- Asian instant soups: <0.01
- Margarine: <0.01
- Biscuits and rice cakes: <0.01

As an example, the calculation of a known moderate diet (1300 kcal) gives the following values for iron intake:

A small breakfast of banana shake containing 150 g of bananas (0.25 mg × 1.5), 200 ml of milk (0.1 mg × 2) and 150 ml of coffee drink (0.2 mg × 1.5) gives a total amount of iron of 0.875 mg.

A good lunch: 200 g fish (1 mg × 2) with 400 g potatoes (0.5 mg × 4), a 200 g apple (0.3 × 2) and 150 ml orange juice (0.04 × 1.5) add up to 4.66 mg iron.

Dinner: 400 g of fruit salad of oranges, kiwis and pineapple (0.3 mg × 4) then give 1.2 mg of iron.

In total, only a rounded 6.7 mg of iron is absorbed. That is only about 50% of the reference value of 14 mg. This is also only the calculation in the best case. The foods used may well contain even less iron than listed here (see iron negative list).

The predominant consumption of iron-poor foods is, among other things, an important reason why iron deficiency exists at all in a healthy person in countries with access to a sufficient food supply – even in very nutrition-conscious people. There is certainly healthy and also varied food on the plate, which also certainly contains all other important macro- and micronutrients, but unfortunately very little iron. So this is quite possible. According to the National Consumption Study II, a high proportion of the population in Germany does not reach the intake recommendations for iron of the German Nutrition Society (see Figs. 9.3 and 9.4).

In order to counteract this, the most important rule is therefore to select foods which contain a relatively safe amount of iron, i.e. where the fluctuation ranges of the iron contents downwards are not so high and which are therefore not included in the iron negative list. They should contain more than 1–2 mg iron/100 g in the ready-to-eat product and should be able to be consumed in larger quantities. If tables of iron contents are used in compiling the diet plan, and there are a large number of these, do not be confused by the indications. If the iron content is based on dry or fresh weight, these indications are even mixed up, or there is no information on this at all.

Every 4 years, the German Nutrition Society (DGE) publishes the Nutrition Report, which presents current data on the nutritional situation in Germany as well as the results of corresponding research projects and serves as a decision-making aid and stimulus for government nutrition and health policy measures (DGE Nutrition Report 2020). The Nutrition Report is published on behalf of and with financial support from the Federal Ministry of Food and Agriculture (BMEL). Today's DGE was founded in 1953 and, according to its statutes, is committed to the common good and science. Today's BMEL was founded in 1949. The first Federal Minister of Agriculture was Prof. Dr. Wilhelm Niklas in the cabinet of Konrad Adenauer. The so-called departmental research of the BMEL includes the Max Rubner Institute (MRI), the Federal Institute for Risk Assessment (BfR) and the German Biomass Research Centre (DBFZ). The tasks of the MRI include the maintenance and further development of the Federal Food Key (BLS). Furthermore, the MRI is responsible for the implementation of National Nutrition Surveys (NVS) on behalf of the BMEL. Within the framework of the NVS II, approx. 20,000 persons between 14 and 80 years of age were interviewed on their dietary behavior between 2005 and 2007. The NVS III will be conducted in the period 2015–2025.

8.3   Examples of Dishes Containing Iron

Here I would like to give you some suggestions for tasty lentil dishes with which you can achieve a good part of your daily iron supply. The classic lentil stew (Fig. 8.28) and lentil hummus with red lentils (Figs. 8.29 and 8.30) and black Beluga lentils (Fig. 8.31). The percentages of iron intake for the individual dishes refer to the daily reference amount of 14 mg.

For two of the dishes presented, the calculation of iron intake is shown exactly under the figure as an example (Figs. 8.28 and 8.30). The mean values from the table for plant foods and the figures from Sect. 8.1 were used. The amount of iron consumed in each portion is shown as a large number next to the finished dish. Figures in the intermediate calculations have been deliberately not rounded so that you can follow the calculation accurately. Taking into account the dry/fresh information – very important – you can quickly make a rough calculation of the iron intake in the same way for each menu you put together yourself.

400 g

Portion

50 %

Fig. 8.28 The iron classic and calculation of iron intake: 500 g of plate lentils, commercially available (40 mg Fe), 200 g of potatoes (1 mg Fe), 200 g of carrots (3 mg Fe), 50 g of peeled lemon (0.15 mg Fe) and 1.5 l of water. Thus, in the total preparation of 2450 g, there are 44.15 mg of iron. By 1/6 of the portion 7.4 mg iron are taken up, which are 53%, thus rounded 50% of the reference quantity of 14 mg

Fig. 8.29 Ingredients for a lentil hummus: red lentils cooked, tahina, garlic, lemon, ground cumin, tomatoes, walnuts, chopped almonds, sunflower seeds. The word hummus comes from the Arabic language and means "chickpea," which is also the basis in the traditional preparation. For better classification in the food category, the name hummus was also used here for the variant with lentils

100 g
Portion
25 %

Fig. 8.30 Delicious, creamy lentil hummus with a good, spreadable consistency. Calculation of iron intake: 300 g red lentils ready to eat (12 mg Fe), 25 g tahina (2.25 mg Fe), 10 g garlic (0.15 mg Fe), 50 g lemon peeled (0.15 mg Fe), 100 g walnuts (4 mg Fe), 50 g almonds (2 mg Fe), 50 g sunflower seeds (2.5 mg Fe), 100 g tomatoes (0.5 mg Fe), 1 g ground cumin (0.7 mg Fe). Thus, in the total preparation of 686 g, there are 24.25 mg of iron. A portion of 100 g provides 3.5 mg of iron, which is 25% of the reference amount of 14 mg

Fig. 8.31 Basic ingredients cooked Beluga lentils, garlic, cumin and lemon (left) and finished hummus after three minutes of processing in a high-powered blender (right)

Recipe for the iron classic for basic care: Lentil stew with carrots and potatoes (serves 6): 500 g lentils, 200 g potatoes, 200 g carrots, 50 g peeled lemon,
1.5 l water, one tablespoon vinegar, one teaspoon pepper, one teaspoon salt.
Boil the lentils in water, cut the potatoes and carrots into small cubes and add
them, cover and continue cooking for about 60 min on low heat, stirring often
and adding water if necessary; only when the lentils and vegetables are cooked
tender, add salt, pepper and the vinegar.

Lentil hummus recipe (serves 6): Cook 250 g red lentils for approx. 12 min,
remove the water through a fine mesh sieve and place 300 g of the ready-to-
eat lentils in a blender, add 25 g tahina (sesame paste), 1 g cumin (ground), a
peeled lemon and two cloves of garlic. Further add to the blender: two tomatoes, 100 g walnuts, 50 g almonds (chopped), 50 g sunflower seeds. Season
with salt and pepper to taste. Blend the ingredients with a powerful blender
for approx. 3 min. and serve the hummus in small portions (Fig. 8.30).

8.4   Rules of Thumb for an Optimal Iron Supply

A healthy person can quite easily meet their daily iron requirements through cleverly composed foods. It is best to use unprocessed ingredients, and if possible from
regional production.

No Predominant Diet with Foods from the Iron Negative List

Of course, the presented iron negative list includes many foods with high-quality
ingredients that the body urgently needs and that are beneficial to health. However,
they unfortunately contain very little iron on average. However, if there are problems with the iron status, the person concerned should not eat predominantly from
an iron-poor food basket, as he or she can then quickly slip into an iron deficiency
despite a supposedly "healthy diet." To reach the reference value of 14 mg iron/day,
for example, 14 kg of apples would have to be consumed, which is relatively impos-
sible. This example shows the content of this first rule of thumb very clearly once
again (cf. also Fig. 8.32).

Additionally Increase the Bioavailability of Iron

Adding lemon juice or other sources of vitamin C to any preparation greatly
increases the absorption of iron, especially from plant foods. For cooked foods, this
should be done after the cooking process (less breakdown of vitamin C by heat).

Fig. 8.32 The factor >10. The iron content of foods in daily use can often differ by more than a factor of 10. Examples include edible peas (2.5 mg Fe/100 g), pineapple (0.25 mg Fe/100 g), and apples (0.1 mg Fe/100 g). This should always be kept in mind when making dietary plans. Decoration: The Blue Dove, Museum Art Christchurch, New Zealand 1991

Red peppers, among others, are particularly rich in vitamin C with over 100 mg/100 g. Do not drink coffee or tea shortly before or after meals. Milk is also not conducive to iron absorption, as the high calcium content hinders absorption in the body. Many small meals spread throughout the day are much more beneficial for iron absorption than one large main meal.

Pulses as a Staple Food

Daily meal should consist of a large proportion of legumes. These include lentils, peas, beans, soybeans. With a consumption of 100 g, based on the dry commercial goods, already half of the daily iron requirement is covered. After swelling and cooking, this quantity results in a ready-to-eat portion of approx. 300–400 g, i.e. a quite normal portion. A large part of the iron in pulses is in the form of ferritin, which is readily bioavailable, and added vitamin C ensures good absorption of the remaining portions. Legumes also have other positive effects. They contain biologically valuable proteins and other minerals, a lot of dietary fiber, they raise the blood sugar level only very slowly and keep it constant for a long time. They are therefore, also independently of their high iron content, an all-round beneficial food.

Cereal Products, Nuts and Seeds on the Menu

Daily should consume 100 g of cereal products, nuts or seeds. These include, for example, oatmeal, pumpkin and sunflower seeds, hazelnuts and walnuts. Individually or in a mixture, you can then give free rein to your imagination when choosing the other ingredients (bananas, apples, pears). In no case should a source of vitamin C be missing. This can easily be achieved by adding an orange smoothie, for example. Milk should not be used. The intake of one third of the daily iron requirement can be easily achieved with it.

Meat Contains More Iron Than Fish

If you only look at the iron content, meat is more beneficial than fish. There are even fish products on the market with extremely little iron (<0.2 mg/100 g). However, eating fish is valuable as a food for many other reasons. In particular, it contains polyunsaturated fatty acids, which play a role in many important bodily functions. Fish should generally not be over fried to preserve the unsaturated fatty acids. Chicken is slightly below beef in average iron levels. Offal should generally be avoided as it is often contaminated with harmful substances. In general, the consumption of meat and fish is not necessary to cover the daily iron requirement.

Green Smoothies Always with Legumes

For in-between and on the go smoothies are very popular and continue to be trendy. In particular, green mixed drinks, which consist of half vegetables/lettuce and half fruit, are of great advantage, as the sugar content is lower than in pure fruit smoothies and the valuable ingredients of vegetables or lettuce can come into their own. Some of the fruit and vegetables can easily be replaced with pulses, e.g. cooked peas, and you get a tasty iron smoothie. The additional use of nuts and seeds is also possible. This then gives the drink a special flavor.

References

Becker S (2007) Inorganic mass spectrometry: principles and applications. Wiley, Chichester

BLS (2021) Bundeslebensmittelschlüssel. https://blsdb.de

DGE (2020) 14. Ernährungsbericht, Deutsche Gesellschaft für Ernährung (Hrsg.) Bonn (ISBN 978–3–88749-269-4)

Eurofir (2021). www.eurofir.org

Klockenkämper R, von Bohlen A (2015) Total-reflection X-ray fluorescence analysis and related methods, 2. Aufl. Wiley, Hoboken

USDA (2021) United States Department of Agriculture. https://www.usda.gov. food data central, https://fdc.nal.usda.gov

Welz B, Sperling M (1999) Atomic absorption spectrometry, 3rd edn. Wiley-VCH, Weinheim

Iron and Special Diets

9

9.1 Vegetarian and Vegan Diet

The vegetarian and vegan lifestyle is in vogue among the population, as is the individualized diet. More and more people are deciding against the consumption of meat or are doing without animal products altogether and are not consuming milk or eggs, for example. There are also other special diets such as frutarians (eat only fruit) or flexitarians (part-time vegetarians). Below you will find an overview of the most important special diets.

Designation and Information on Different Diets

Omnivore: eats a variety of food of both plant and animal origin.

Vegetarians: Upper group, do without the consumption of animal products as far as possible, there are the different subgroups, vegans, ovo-lacto-vegetarians, ovo-vegetarians, lacto-vegetarians, pescetarians, frutarians, flexitarians and "pudding-vegetarians."

Vegetarians consume besides the vegetable food – if – then only the products of the living animal (e.g. egg, milk, honey) and from it manufactured food and additives, no slaughter by-products (e.g. collagen, lard) or animal gelatine (alternative Agar-Agar from algae, also called Chinese or Japanese gelatine).

Vegans: Eat only plant foods, all animal products such as meat, fish including seafood, eggs, milk and honey are not consumed.

Ovo-lacto-vegetarians: In addition to the plant-based diet, meat, fish and seafood are not consumed, eggs, milk and their processed products are eaten.

Ovo-vegetarian: diet as for ovo-lacto-vegetarian, but additional renunciation of milk and dairy products (can be replaced by soy products).

Lacto-vegetarian: diet as for the ovo-lacto-vegetarian, but additional renunciation of eggs and egg products.

K. Günther, *Diet for Iron Deficiency*, https://doi.org/10.1007/978-3-662-65608-2_9

Pescetarians: No consumption of meat of warm-blooded animals, fish and mostly sea animals are allowed.

Frutarians: nutrition is exclusively based on plant products that do not damage the parent plant, e.g. fruit, seeds or nuts.

Flexitarians: So-called part-time vegetarians consume meat or fish only once or twice a week and eat mainly plant-based foods.

Pudding vegetarians: Joking term for people who abstain from meat consumption but otherwise do not pay much attention to healthy nutrition, frequent consumption of convenience foods and foods with few valuable substances.

Mediterranean diet: In this currently very popular diet, a lot of fish is consumed, which is considered very valuable, especially because of the often high proportion of polyunsaturated fatty acids. In addition, fruit, vegetables and salad are part of the daily menu. The consumption of red meat is low.

People with lactose intolerance: the breakdown of milk sugar is impaired, leading to abdominal pain and flatulence, severe restriction of the consumption of milk and dairy products such as cream or yoghurt.

Persons with celiac disease: gluten intolerance due to a chronic disease of the small intestine; gluten is a protein mixture which is found in wheat, barley, green spelt and spelt, among others.

Halal diet: practiced by devout Muslims, all plant-based foods are halal, exceptions are products that have an intoxicating effect or are toxic (nutmeg, alcohol); pork and pork products (e.g. gelatine) are generally prohibited, other meat is only permitted if the animal has been ritually slaughtered.

Kosher diet: It is practiced by observant Jews and divides foods into meaty, dairy and neutral (e.g. fruits, vegetables, grains). Neutral foods may be eaten together with meaty and milky foods. Fish may not be eaten together with meat, and dairy products and meat may not be eaten together or even prepared, pork meat and products are generally forbidden, other meat is only allowed if the animal has been ritually slaughtered, consumption of ruminants with cloven hooves (beef, deer, goat, sheep) and fish with scales and fins (trout, salmon, tuna) is allowed.

Low carb diet: Mainly used to reduce weight, the intake of carbohydrates is severely restricted, especially sugar. Protein and fat may be consumed, if cereal products are on the menu, then only as whole grains.

Separation diet: mainly used for weight reduction, strict separation of foods with protein (meat, fish, milk, eggs) and carbohydrates (potatoes, rice, pasta, sugar, cereals, bananas); the background for this special diet is the assumption that the digestive tract can adjust better to the utilization of the different ingredients, between the consumption of carbohydrates and protein should be about 3 h.

Paleo or Stone Age diet: Industrially processed foods are avoided, as well as alcohol, milk, sugar, but also grains and legumes. Only what our ancestors from the Stone Age (end about 5000 years ago) have eaten is consumed: Meat, fish, seafood, vegetables, eggs, fruit, nuts.

Raw foodists: All foods that have been heated are avoided, e.g. coffee or tea, roasted nuts, roasted meat or cooked vegetables.

There are about 1 billion vegetarians worldwide, and the reasons for this are manifold. In addition to arguments relating to certain food ingredients, ethical considerations and animal welfare, but also the high energy expenditure in the production of animal foods and climate protection are cited.

The vegetarian lifestyle is not a new phenomenon, however, because this particular attitude to nutrition has existed for a long time. Well-known vegetarians or vegans include Pythagoras (philosopher, around 570–500 BC.), Leonardo da Vinci (inventor and painter, 1452–1519), Wilhelm Busch (poet, 1832–1908), George Harrison (musician, 1943–2001), Al Gore (politician and Nobel Peace Prize winner, *1948), Christoph Maria Herbst (actor, *1966), Leonardo DiCaprio (actor, *1974) and Lewis Hamilton (racing driver, *1985) (Leitzmann and Keller 2020).

The leading organization in the field of plant-based nutrition, ProVeg-International (proveg.com/en), estimates that in 2021 the proportion of vegetarians in Germany will be around 8 million, which corresponds to 10% of the population. Furthermore, 1–2% eat a vegan diet. The trend for both groups continues to rise. The food trade has reacted to this trend with a growing proportion of veggie products in its range, and plant-based alternatives can now be found in every supermarket.

Studies have shown that many more women are vegetarian or vegan than men. Furthermore, their proportion is particularly high among young women. The proportion of vegetarians in different countries is partly very different. In India, for example, almost 40% of people are vegetarians, whereas in Australia this figure is only 5%.

Vegetarians have on average a lower body weight and a lower body mass index (BMI) compared to omnivores (mixed diet) (Fig. 9.1). Obesity is very rarely found in vegans in particular, and plant-based diets therefore have considerable potential in the therapy of obese patients (Leitzmann and Keller 2020).

In a position paper of the German Nutrition Society, various nutrients are described as critical for the vegan lifestyle (Fig. 9.2). This does not mean, however, that one automatically develops a deficiency state with a purely plant-based diet, but that one can certainly avoid the undersupply of nine of the factors shown in Fig. 9.2 by consuming a skillful combination of different foods. This will be discussed later in relation to iron.

An exception is vitamin B_{12}, which should be supplemented by a vegan diet according to the current state of science (see also Sect. 4.2).

But there are also critical nutrients in the diet of the general population, as surveys of the National Nutrition Survey II (NVS II) show (Figs. 9.3 and 9.4). For example, in the 14 to 50 age group of female study participants, 75% do not reach the reference value for iron. This is a considerable proportion and, against the background of the many important functions of the element in the organism, a fact to which more attention urgently needs to be paid.

The essential character of iron in the synthesis of the important neurotransmitters and hormones, dopamine, norepinephrine, epinephrine, serotonin and melatonin and the connection with depressive moods has already been pointed out in Sect. 2.6. Against the background of the currently publicly discussed increase in depressive

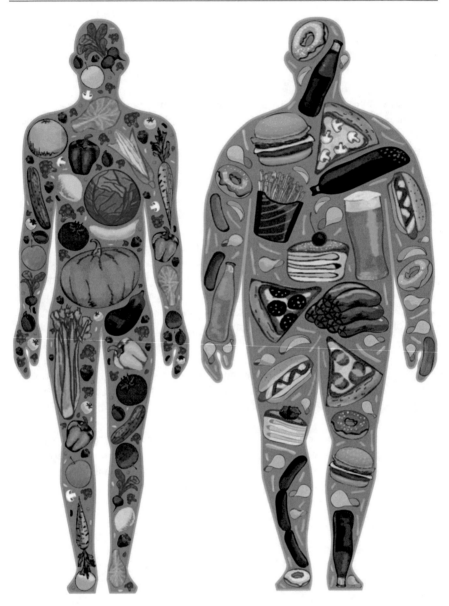

Fig. 9.1 The choice of which foods to eat is, of course, individual and free. Studies show that a plant-based diet has a lower proportion of people with obesity. (© Designincolor/stock.adobe.com)

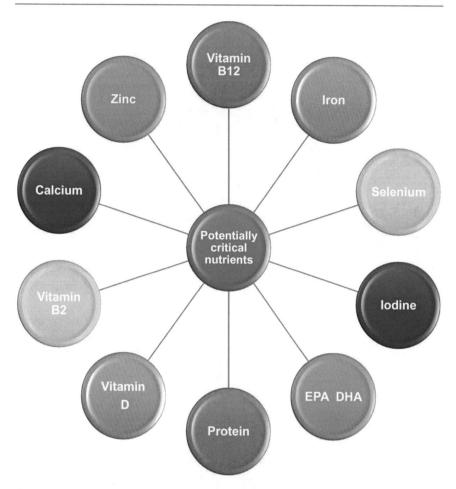

Fig. 9.2   Potentially critical nutrients in a vegan diet according to the position paper of the German Nutrition Society (Richter et al. 2016)

disorders in the population, this is certainly information that should be given much more attention.

The National Nutrition Survey III (NVS III) will be conducted from 2015 to 2025. The executing institution is the Max Rubner Institute in Karlsruhe, funding is provided by the Federal Ministry of Food and Agriculture (BMEL), and the executing agency is the Federal Agency for Agriculture and Food (BLE) in Bonn. If the NVS III confirms the result of the NVS II that three quarters of the female study participants do not reach the reference value for iron, one should certainly consider initiating new educational campaigns on this point for the population, but also for the advisory professional contacts, or to further strengthen already existing ones.

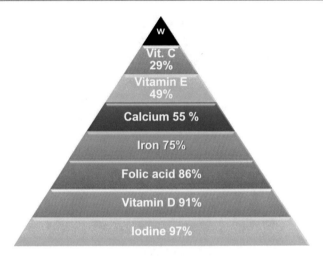

Fig. 9.3 Critical nutrients in Germany according to data from the National Nutrition Survey II (MRI 2008). The percentages are the proportion of affected women who do not reach the reference value. Iodine: if iodized table salt is not used. Iron: Age group 14–50 years

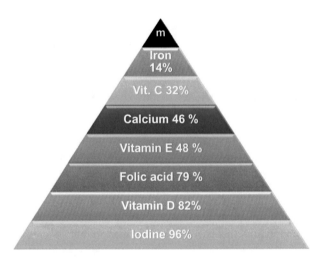

Fig. 9.4 Critical nutrients in Germany according to data from the National Nutrition Survey II (MRI 2008). The percentages are the proportion of affected men who do not reach the reference value. Iodine: if iodized table salt is not used

Information on Vitamin B_{12} (Cobalamin) (Leitzmann and Keller 2020) (cf. also Fig. 4.4).

Cobalamins cannot be produced by plants, and therefore vitamin B_{12} is the most critical micronutrient in a purely plant-based diet. Cobalamins are produced exclusively by microorganisms through biosynthesis, and several animal species meet their needs through the self-synthesis of the gastrointestinal microbiome. Humans cannot use the cobalamin produced by the colon microbiome because absorption occurs in the terminal illeum (last section of the small intestine), and therefore rely on dietary intake.

In the human organism there are two important vitamin B_{12}-dependent biochemical reactions: Adenosylcobalamin is cofactor of methylmalonyl-CoA mutase (MCM, fatty acid and amino acid degradation) and methylcobalamin is required by methionine synthase to catalyze the remethylation of homocysteine to methionine, which also provides a link to folate metabolism (Fig. 4.5).

The European Food Safety Authority (EFSA) set the Adequate Intake (AI) of vitamin B_{12} at 4 µg/day for adults and the D-A-CH societies raised the reference value from 3 to 4 µg/day for adults in 2019. The vitamin B_{12} content of animal foods varies widely. High levels of >2 µg/100 g are found in e.g. beef liver (65 µg/100 g), herring (11 µg/100 g), beef (5 µg/100 g) or camembert (3 µg/100 g). In the medium range of 0.3–2 µg/100 g are e.g. chicken eggs (2 µg/100 g), redfish (1.5 µg/100 g) or cow's milk (0.4 µg/100 g).

Due to the gastric acid, the vitamin B_{12} ingested with the food is released from protein complexes and finally binds to the intrinsic factor (IF), a glycoprotein formed in the occupant cells of the gastric mucosa. This B_{12}-IF complex is then absorbed in the ileum. Independently of this, vitamin B_{12} can also be absorbed via passive diffusion without IF across the mucosa of the gastrointestinal tract and oral cavity, but this only occurs at high concentrations. However, the absorption rate by this route is only a maximum of 1%.

Due to the still used outdated microbial measurement methods for the determination of vitamin B_{12} in food, the content of the really active form is often overestimated, as inactive B_{12} analogues are also determined. Modern mass spectrometry methods, on the other hand, can distinguish between active and inactive forms of B_{12} and should be used more in the future.

So far, there are no reliable findings on the physiological activity of vitamin B_{12} in the algae preparations that are sometimes offered on the market. Therefore, these preparations should not be used for B_{12} supplementation under any circumstances at present. Chlorella and spirulina, which are usually available in the form of dietary supplements, are the most important of these. In this area, however, there are currently interesting research approaches that could possibly lead to plant-based preparations with bioactive vitamin B_{12} in the future.

Contrary to widespread belief, vegetarians do not have an iron status that is detrimental to health (Haider et al. 2018) if they pay attention to a targeted selection and composition of their diet, and the increased consumption of red meat is not assessed as beneficial (Ekmekcioglu et al. 2018; DGE 2020). The surprisingly good iron supply of vegetarians is attributed to various reasons, and the newly discovered uptake pathway of plant ferritin could certainly play a major role here (cf. Sect. 7.3).

Some scientists are now of the opinion that the bioavailability of plant iron, i.e. the proportion that is actually absorbed by the body, is just as good as that of iron from animal products. However, the entry into the cells is said to be slower, more uniform and, above all, gentler (Theil 2017; Zielinska-Dawidziak 2015; Theil et al. 2012). In the future, this will open up interesting possibilities for the development of new stomach-friendly drugs based on plant ferritin (phytoferritin).

When looking at the iron content as a whole, there is also no disadvantage for plant products, it is in some cases even much higher, as you could already see in Chap. 8. This means that vegetarians and vegans simply have to make sure that they give adequate consideration to plant products with a sufficient iron content in their daily food selection.

As already shown in Sect. 8.3, a good portion of pulses, such as lentils, is sufficient to cover about half of the daily iron requirement. If you then make sure that you do not eat exclusively foods from the "iron negative list" (see Sect. 8.2) for a long time, you should no longer have any problems with your iron supply. This is good news for all people who eat a vegetarian or vegan diet.

Frutarians, on the other hand, can easily slip into an iron deficiency because very little of the trace element is present in many of the foods consumed by this group. The same applies to the people who are often jokingly called "pudding vegetarians." With the rest of the diets presented, it is no problem to ensure iron supply through a targeted selection of plant and animal foods.

As already emphasized several times, a good portion of legumes per day is sufficient to cover about 50% of the daily iron requirement. Lentils, chickpeas, as well as legumes in general, should only be consumed when cooked. When raw, they contain lectins (lentils) or phasins (chickpeas), which may have a negative effect on the organism.

In addition to the high iron values, the other valuable constituents of legumes are also impressive. The summary in Table 9.1 shows the typical composition of beluga lentils, which are so called in reference to beluga caviar, the roe of the beluga, the largest sturgeon species (Fig. 9.5).

In addition to over 40% carbohydrates, the high contents of protein and fiber are particularly noticeable. With 30%, lentils are very rich in protein with a high lysine content and with 15% fiber content very beneficial for the diet. Although dietary fiber has no nutritional value, it is extremely important for digestion and provides a long-lasting feeling of satiety.

Important vitamins in lentils are in particular thiamine (vitamin B_1), which is important for the nervous system, and folic acid, which has already been discussed and is necessary for blood formation, among other things.

Table 9.1 Average nutritional values of commercially available Beluga lentils per 100 g uncooked (manufacturer's data, rounded), the percentages correspond to the proportions of the nutrient reference value, i.e. the recommended daily intake

Energy	1400 kJ/335 kcal	% NRV
Fat	1.4 g	
Of which saturated fatty acids	0.3 g	
Carbohydrates	44 g	
Of which sugar	0.7 g	
Fiber	15 g	
Protein	30 g	
Salt	<0.01 g	
Phosphorus	450 mg	65%
Thiamine	0.4 mg	36%
Folic acid	120 µg	62%
Iron	7.0 mg	50%
Magnesium	120 mg	32%
Zinc	4 mg	40%

Fig. 9.5 Beluga lentils as dried commercial goods

Fig. 9.6 The iron classic with Beluga lentils (see Sect. 8.3). When this dish is eaten, approx. 50% of the reference amount of iron of 14 mg is absorbed. The potatoes are colored black-blue by the value-giving bioflavonoids and thus there is a further health advantage when eating this type of lentil

Magnesium and zinc are important for the function of many enzymes that play a major role in energy and hormone metabolism. Zinc is the second most common trace element in the human body after iron and is also very important for the effect of insulin. A zinc deficiency often leads to wound healing disorders and increased susceptibility to infections.

When eating 100 g of Beluga lentils (dry, commercial goods), which corresponds to a well-filled plate after cooking, one has already consumed 7 mg of iron, i.e. 50% of the reference value of 14 mg, according to Table 9.1. With this simple rule of thumb in mind, you can always cover half of your daily requirement very quickly, even on a vegan and vegetarian diet. The iron values of the other lentil varieties are also in this range (5–10 mg iron/100 g commercial goods), so that one is also independent of the variety.

Figure 9.6 shows the tried and tested basic dish (cf. Fig. 8.28) with beluga lentils, so that you also get a visual impression. Here it is noticeable that the natural black dye of the lentils also darkens the other menu components when processed together, giving the dish an interesting visual touch.

These blue-black, violet or even red pigments, also called anthocyanins, are also found in other foods such as blackberries, black currants, cherries or raspberries. These secondary plant substances belong to the bioflavonoids and polyphenols. Many positive effects on the organism are attributed to them or their degradation products, e.g. the destruction of free radicals, which are often responsible for cell damage (see Sect. 6.3). However, anthocyanins are also involved in many other important protective mechanisms in the body and are the subject of numerous

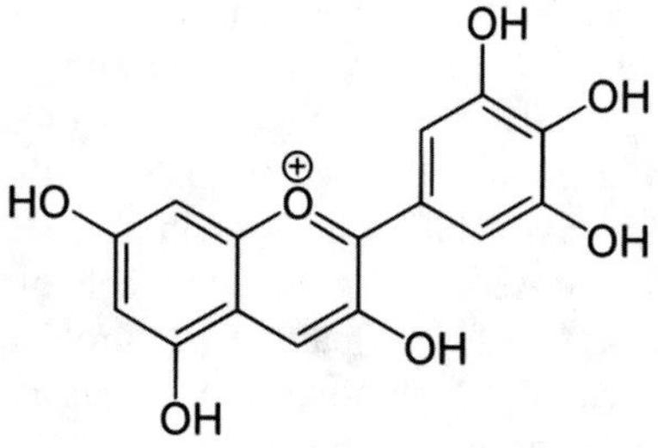

Fig. 9.7 Delphinidin cation, an important member of the anthocyanin group and a value-giving constituent of, among other things, Beluga lentils. The color of anthocyanins changes with the pH value and upon complexation

research projects. Among other things, anti-carcinogenic effects and positive effects on inflammation and neurodegenerative diseases are being discussed (Li et al. 2017).

A delphinidin derivative is responsible for the black color of the Beluga lenses. The structure was elucidated by liquid chromatography, mass spectrometry and 1- or 2-dimensional nuclear magnetic resonance spectroscopy. It is delphinidin linked to glucose and arabinose (Takeoka et al. 2005). Figure 9.7 shows the basic structure of delphinidin. The phenolic OH groups are important for the free radical scavenging process.

9.2   Childhood, Youth and Old Age

An optimal supply of the trace element iron is very important, especially in the various childhood growth phases. In addition to physical development, iron is absolutely essential for later cognitive performance and also for the social maturation process. Numerous important enzymes and brain processes are controlled by the trace element. Among other things, it is involved in the formation of connections between nerve cells or the production of important neurotransmitters. These neurotransmitters have an enormous impact on general brain function and also emotional states, as already discussed in Sect. 2.6 (Hect et al. 2018; Kim and Wessling-Resnick 2014). Even iron deficiency without anemia (see Sect. 5.1) can be the cause of low mood, tension and fatigue (Sawada et al. 2014).

Furthermore, studies show that iron homeostasis is different in infants due to the lack of regulation of the divalent metal transporter (DMT-1) or ferroportin (cf. Sect. 3.1) and that special physiological conditions therefore exist here (Lönnerdal 2017).

Numerous diseases such as ADHD (attention deficit hyperactivity disorder), autism, mood swings and developmental disorders can be attributed in part to iron deficiency in children (Chen et al. 2013). ADHD is characterized by motor restlessness, attention deficit disorder, and impulsivity. In children with ADHD, before considering therapy with psychostimulants (e.g. Ritalin), nutritional habits should also be kept in mind. Perhaps this is the cause of the problem. An iron-deficient diet is also possible in children if the daily meals are poorly composed (cf. iron-negative list, Sect. 8.2).

The same applies to teenagers and young adults who eat an unbalanced, low-calorie diet following the slimming trend. If the foods are combined incorrectly, especially over a longer period of time, an iron deficiency is inevitable. Those affected are then often pale, without energy and show increased irritability. Especially during the growth phase, an incorrect diet can quickly lead to iron deficiency. About 1% of all adolescent girls suffer from eating disorders, also known as anorexia. This is particularly problematic for the iron supply, also because menstruation, which often starts at this age, makes the situation even worse.

According to Table 6.3, the reference values for the daily iron intake of the D-A-CH societies for children aged 1–7 years are 8 mg and from the age of 7 years already 10 mg, which is in the order of magnitude of adults and then does not increase so much. In the range 10–19 years, reference values of 15 mg/day (female) and 12 mg/day (male) are given. For infants, the reference values for daily iron intake of the D-A-CH societies are 0.5 mg (0–4 months) and 8 mg (4–12 months) (Table 6.4).

The comparison of the reference values of the nine different societies and organizations considered for infants, children and adolescents shows major differences in some areas (cf. Tables 6.3 and 6.4). As an example, the very small reference value of 0.2 mg iron/day for infants (0–6 months) of the National Health and Medical Research Council (Australia and New Zealand) should be highlighted here, whereas other organizations state a multiple of this amount, of up to 5 mg/day (0–6 months, Health Council, NL, cf. Table 6.4).

The worldwide harmonization of reference values for iron is therefore an important, but certainly not easy task for the professional societies in the future, whereby the latest scientific findings in all fields of iron biochemistry must be taken into account (cf. also Chap. 6).

The "iron hunger" of children and adolescents can be well satisfied by a clever composition of food. For children, porridges with oat or millet flakes, which have a high iron content, are particularly suitable (see Sect. 8.1). Older children, on the other hand, can also prevent iron deficiency by eating pulses such as peas, beans or lentils, or, if they wish, eat a meal with meat more often.

For young people, legumes are often "old-fashioned food" and therefore usually not popular, or are not consumed according to contemporary peer pressure. Smoothies with pulses, for example, can provide a remedy here. A homemade green smoothie typically consists of 50:50 fruit and lettuce. Unfortunately, fruit and lettuce are usually not very rich in iron (cf. Chap. 8).

In contrast, legumes and also grain products are much richer in iron and can be used in the preparation of smoothies. This gives you a quick and convenient "hip iron booster" that is easily transportable and can be conveniently consumed anywhere. The high fiber content and the valuable proteins in legumes also provide a long-lasting feeling of satiety and a healthy diet. So perfect for the care of the slim figure, to prepare the body for the working day, strenuous exams or the upcoming party.

People in their more mature years often eat less, and so the iron supply can suffer. In addition, the intestine can no longer utilize nutrients as well, and the utilization of micronutrients such as iron is poorer. Furthermore, the risk of small bleedings in the digestive tract increases and with it the loss of iron. Increased intake of medications can also lead to a decrease in iron absorption, as they can bind the trace element. Medications are often iron robbers.

In old age, iron deficiency is therefore relatively common, and therefore the status should be checked more often by a doctor. Since inflammatory processes are more frequently present in the body in this group of people, the ferritin value is often not useful as a marker. Therefore, more comprehensive, individual-dependent parameters are necessary to clarify the iron status in older persons, such as the determination of hepcidin, the soluble transferrin receptor or the performance of other additional examinations (Burton et al. 2020; Joosten 2018).

However, a deficit can be successfully counteracted by an iron-rich diet. Natural, iron-rich foods such as lentils, beans or wholemeal products, also in pureed form, can – as in the other age groups – contribute to enjoying a high level of physical and mental fitness at an advanced age. Especially in people who experience or have experienced this happiness, it is noticeable that a conscious diet in particular, in addition to regular exercise and mental activity, have contributed significantly to this.

9.3 Pregnancy and Breastfeeding Period

From a single fertilized egg cell, the embryo forms via the 2, 4, etc. cell stages (embryogenesis). After about 9 weeks, the embryo develops into the fetus, called fetogenesis, which finally ends with birth. After that, the breastfeeding period often begins, which in Germany is on average about half a year.

All these biological processes from the fertilized egg cell to the adult human being require a high temporal and spatial order and the interaction of many biochemical processes, in which iron is also strongly involved. Therefore, especially in the stage of development, the optimal supply of nutrients for mother and child is particularly important so that no health disadvantages occur (Benson et al. 2021; Iglesias 2018). In pregnancy and during the breastfeeding period, several organisms must be supplied with the vital trace element, and therefore the demand is much higher.

The daily iron intake recommended by the D-A-CH societies for pregnant and lactating women is given as 30 mg and 20 mg/day, respectively, whereas EFSA lists reference values of 16 mg/day for both groups (cf. Table 6.2). These widely differing recommendations result from the different derivations of the two organizations and are probably mainly due to the assumption of a higher iron bioavailability in the EFSA approach. In Sect. 6.1, these circumstances were discussed in more detail and the EFSA derivation for pregnant women was presented.

The reference value of the D-A-CH societies for pregnant women is thus twice as high as for women up to approx. 50 years of age with 15 mg/day (cf. Table 6.1). The recommended 30 mg/day can certainly be achieved by a good combination of different foods, as a rough calculation shows:

100 g mixture (50:50) of wheat bran (16 mg/100 g) and pumpkin seeds (12 mg/100 g),
the mixture then has 14 mg iron/100 g, with 150 ml freshly squeezed orange
juice (0.04 mg × 1.5) and 150 g banana (0.3 mg × 1.5) = approx. 14.5 mg iron
One serving of lentil stew = 7.5 mg iron
50 g nuts for snacking (8 mg × 0.5) = 4 mg iron
200 g meat (3 mg × 2) = 6 mg iron
This results in a total of 32 mg of iron, and you have reached the daily recommended amount for pregnant women of the D-A-CH societies. For vegetarians and
vegans, meat can easily be replaced by vegetable products with a similarly high iron
content. Furthermore, one can still replenish one's iron supply by treating oneself to
something tasty or sweet and snacking on dark chocolate (Fig. 8.21) or using sugar
beet syrup to sweeten food instead of household sugar (Fig. 8.22).

Of course, this is only a suggestion for a daily food selection, which is only
intended as an example to show how, by cleverly putting together a diet plan according to the iron contents normally found in foods (cf. Chap. 8), one can quite possibly
reach 30 mg iron intake per day.

9.4   Sports Activities

People who are very physically active or even do competitive sports generally have
an increased basal metabolic rate. This also increases the need for vital substances
and thus also for essential trace elements such as iron. If there is a deficiency, this is
a major obstacle to the achievement of athletic performance or even top performance. Sport is iron murder, as the saying goes, and there is some truth to it.

In addition to losses via sweat, the cause of the increased need for iron during
sporting activities is the increased muscle build-up (myoglobin, iron-containing
enzymes), the higher total blood volume (more erythrocytes and hemoglobin), but
also micro-bleeding, which can occur in various parts of the body during heavy
exertion. They may cause a higher iron loss than the other factors mentioned, which
intensive athletes in particular should be aware of.

The iron requirements of female athletes are reported to be 70% higher than
those of non-athletes, and it is possible that non-anemic iron deficiency is also detrimental to athletic performance. The mean concentration of iron in sweat at 60 min
of exercise is 0.56 mg/l. For a 70 kg person at an exercise intensity of 10 km/h (running, 15 °C outdoor temperature), this results in a sweat rate of 0.8 l/h (Carlsohn
et al. 2019).

During a training session of 1 h, 0.45 mg of iron is excreted via sweat. Assuming
an iron absorption rate of 10%, 4.5 mg of iron would have to be returned to the body.
This is almost 50% of the D-A-CH reference value for men and about one third of
the corresponding value for women (19–51 years) (cf. Table 6.1).

In this context, questions about the bioavailability of iron during and after physical activity are naturally also very interesting. In related studies the focus is particularly on the influence of physical activity on the concentration of hepcidin, as it is

considered the central regulator of iron metabolism (Sim et al. 2019) (cf. Sect. 3.3). Resting levels of key iron parameters play a dominant role in the regulation of hepcidin in elite athletes following endurance training (Peeling et al. 2017), and resistance and endurance training caused a significant increase in serum concentrations of hepcidin and interleukin-6 (IL-6, affects hepcidin levels, Fig. 3.9) after exercise (Larsuphrom and Latunde-Dada 2021), and thus iron release into the blood circulation would be inhibited (cf. Sect. 3.3).

A final evaluation of the complex relationships between iron loss, increased consumption and altered bioavailability to derive practical recommendations for the sports sector is currently still pending.

Risk Groups for Iron Deficiency in People with Sporting Activities
Athletes and especially female athletes with low body weight. In many sports, body weight is greatly reduced by special diets in order to create better conditions for competition. If the diet consists mainly of iron-deficient foods (cf. Sect. 8.2), iron deficiency and the associated drop in performance are inevitable. The advantages of lower body weight are therefore negated.

Extreme athletes, such as marathon runners. Micro-bleeding in the parts of the body that are particularly stressed can lead to severe iron loss, e.g. pendulum movements during running.

Endurance athletes who mainly eat a diet rich in carbohydrates. These often contain very little iron and are also poor in other micronutrients. Unfortunately, this is often overlooked in this type of diet.

Athletes under pain therapy with so-called NSAIDs, non-steroidal anti-inflammatory drugs. NSAIDs are anti-inflammatory and analgesic drugs that are not derived from sterols (basic structure of hormones). Among other things, they prevent the formation of prostaglandins. As a result, the pain and inflammatory processes mediated by this group of substances decrease. NSAIDs can trigger mucosal defects in the gastrointestinal tract as a side effect, leading to gradual blood loss.

All persons who are very active in sports should have their iron status checked frequently by a doctor or sports physician by means of meaningful blood values (ferritin, transferrin saturation, cf. Chap. 5) and, if an iron deficiency is detected, obtain information there about preparations for sportsmen and sportswomen. Independent iron supplementation without medical supervision is strongly discouraged.

Through cleverly selected food combinations, however, athletes can create a good foundation that prevents iron deficiency. For example, a good portion of legumes before the competition or during the training phases is often more helpful and sustainable than many of the bars or fitness drinks offered today.

Sport is also playing an increasingly important role in the general population, and therefore questions around proper nutrition and beneficial foods for physical activity are now very common to professionals such as nutritionists and food scientists. In these consultations, current textbooks with a clear compilation of the whole topic can support very well (Jeukendrup and Gleeson 2019; Lamprecht et al. 2017).

References

Benson C, Shah A, Stanworth S, Frise C, Spiby H, Lax S, Murray J, Klein A (2021) The effect of iron deficiency and anaemia on women's health. Anaesthesia 76:84–95

Burton J, Yates L, Whyte L, Fitzsimons E, Stott D (2020) New horizons in iron deficiency anaemia in older adults. Age Ageing 49:309–318

Carlsohn A, Braun H, Großhauser M, König D, Lampen A, Mosler S, Nieß A, Oberritter H, Schäbethal K, Schek A, Stehle P, Virmani K, Ziegenhagen R, Heseker H (2019) Minerals and vitamins in sports nutrition. Position of the working group sports nutrition of the German Nutrition Society (DGE). Ernährungs-Umschau 66:250–257

Chen M, Su T, Chen Y, Hsu J, Huang K, Chang W, Chen T, Bai Y (2013) Association between psychiatric disorders and iron deficiency anemia among children and adolescents: a nationwide population-based study. BioMed Central (BMC) Psychiatry 13:161

DGE (2020) 14. Ernährungsbericht der Deutschen Gesellschaft für Ernährung. Abschn. 5.1: Gemüse-, Obst- und Fleischverzehr und das Risiko für ausgewählte ernährungsmitbedingte Erkrankungen: Ein Umbrella Review von Metaanalysen

Ekmekcioglu C, Wallner P, Kundi M, Weisz U, Haas W, Hutter H (2018) Red meat, diseases, and healthy alternatives: a critical review. Crit Rev Food Sci Nutr 58:247–261

Haider L, Schwingshackl L, Hoffmann G, Ekmekcioglu C (2018) The effect of vegetarian diets on iron status in adults: a systematic review and meta-analysis. Crit Rev Food Sci Nutr 58:1359–1374

Hect J, Daugherty A, Hermez K, Thomason M (2018) Developmental variation in regional brain iron and its relation to cognitive functions in childhood. Dev Cogn Neurosci 34:18–26

Iglesias L (2018) Effects of prenatal iron status on child neurodevelopment and behaviour: a systematic review. Crit Rev Food Sci Nutr 58:1604–1614

Jeukendrup A, Gleeson M (2019) Sport nutrition, 3rd edn. Human Kinetics, Champaign

Joosten E (2018) Iron deficiency anemia in older adults: a review. Geriatr Gerontol Int 18:373–379

Kim J, Wessling-Resnick M (2014) Iron and mechanisms of emotional behaviour. J Nutr Biochem 25:1101–1107

Lamprecht M, Holasek S, Konrad M, Seebauer W, Hiller-Baumgartner D (Hrsg) (2017) Lehrbuch der Sporternährung, 1. Aufl. CLAX-Fachverlag, Graz

Larsuphrom P, Latunde-Dada G (2021) Association of Serum Hepcidin Levels with aerobic and resistance exercise: a systematic review. Nutrients 13:393. https://doi.org/10.3390/nu13020393

Leitzmann K, Keller M (2020) Vegetarische und vegane Ernährung, 4. Aufl. Eugen Ulmer KG, Stuttgart

Li D, Wang P, Luo Y, Zhao M, Chen F (2017) Health benefits of anthocyanins and molecular mechanisms: update from recent decade. Crit Rev Food Sci Nutr 57:1729–1741

Lönnerdal B (2017) Development of iron homeostasis in infants and young children. Am J Clin Nutr 106:1575–1580

MRI (2008) Nationale Verzehrsstudie II, Ergebnisbericht Teil 2. Max Rubner-Institut, Karlsruhe

Peeling P, McKay A, Pyne D, Guelfi K, McCormick R, Laarakkers C, Swinkels D, Garvican-Lewis L, Ross M, Sharma A, Leckey J, Burke L (2017) Factors influencing the post-exercise hepcidin-25 response in elite athletes. Eur J Appl Physiol 117:1233–1239. https://doi.org/10.1007/s00421-017-3611-3. (PMID: 28409396)

Richter M, Boeing H, Grünewald-Funk D, Heseker H, Krohe A, Leschik-Bonnet E, Oberritter H, Strohm D, Watzl B (2016) Vegane Ernährung. Position der Deutschen Gesellschaft für Ernährung e. V. (DGE). Ernährungsumschau 63:92–102

Sawada T, Konomi A, Yokoi K (2014) Iron deficiency without anemia is associated with anger and fatigue in young Japanese woman. Biol Trace Elem Res 159:22–31

Sim M, Garvican-Lewis L, Cox G, Govus A, MacKay A, Stellingwerf T, Peeling P (2019) Iron considerations for the athlete: a narrative review. Eur J Appl Physiol 119:1463–1478

Takeoka G, Dao L, Tamura H, Harden L (2005) Delphinidin-3-O-(2-O-beta-D-Glucopyranosyl-alpha-L-arabinopyranoside): a novel anthocyanin identified in Beluga black lentils. J Agric Food Chem 53:4932–4937

Theil EC (2017) Methods for isolation, use and analysis of ferritin. United States patent application publication no.: US2017/0087209A1

Theil E, Chen H, Miranda C, Janser H, Elsenhans B, Nunez M, Pizarro F, Schümann K (2012) Absorption of iron from ferritin is independent of heme iron and ferrous salts in woman and rat intestinal segments. J Nutr 142:478–483

Zielinska-Dawidziak M (2015) Plant ferritin – a source of iron to prevent its deficiency. Nutrients 7:1184–1201

Iron Supplementation

10

10.1 Medicinal Products

Medicinal products are substances and preparations which are intended to cure or alleviate diseases or which are used to prevent diseases from occurring in the first place. The legal definition is described in § 2 of the German Medicines Act (AMG), the purpose of which is to ensure the safety of medicines in particular.

Modern finished medicinal products as defined by the AMG may only be placed on the market if they have a German or European marketing authorization. The approval and registration of medicinal products in Germany is carried out by the Federal Institute for Drugs and Medical Devices (BfArM) in Bonn. The approval is granted for a limited period of 5 years. After this period, the benefits and risks are re-evaluated. Nowadays, medicinal products are therefore very well monitored and controlled products.

The preferred, first way for iron deficiency is oral therapy. There are many preparations in tablet, juice or effervescent form on the market. They mostly contain compounds with Fe-(II), but Fe-(III) preparations are also available. They differ in dosage and in the chemical form in which the iron ions are present. This can strongly influence the bioavailability.

A major challenge with the oral supply of iron is, among other things, the poor gastric tolerance of many preparations. Some time after intake, patients often report stomach discomfort or nausea, especially after higher doses. As a result, many patients discontinue therapy.

The amount of iron per intake unit for severe iron deficiency is between 25 and 100 mg. Iron administration must usually be carried out over several weeks to months. Then first the deficit in the blood is balanced and finally the iron store is replenished. Many iron preparations for oral therapy do not require a prescription, but are available from pharmacies.

© The Author(s), under exclusive license to Springer-Verlag GmbH, DE, part of Springer Nature 2023
K. Günther, *Diet for Iron Deficiency*,
https://doi.org/10.1007/978-3-662-65608-2_10

Weekly iron and folic acid supplementation is among the top eight effective global interventions to improve adolescent nutrition and iron status identified by WHO in the 2018 guidelines (Roche et al. 2021).

Currently Available Oral Iron (II) and Iron (III) Preparations (Modified According to Farrag et al. 2019)

- Iron-(II)-sulfate, most commonly used, very inexpensive
- Iron-(II)-bisglycinate, complex with the simplest amino acid glycine
- Iron-(II)-fumarate, complex with the dicarboxylic acid fumaric acid
- Iron-(II)-gluconate, complexed with the monocarboxylic acid gluconic acid
- Iron-(II)-citrate, complexed with the hydroxytricarboxylic acid Citric acid
- Iron-(III)-polymaltose, complex of ferric hydroxide and polymaltose; maltose is a disaccharide of two glucose molecules and a degradation product of starch
- Iron-(III)-maltol, complex with three molecules of maltol (3-hydroxy-2-methyl-4-pyrone), approval in the EU 2016, use in inflammatory bowel disease.

The solubility of Fe-(II) ions only decreases from pH > 6, whereas Fe-(III) ions are only stable in a strongly acidic environment and hydrolyse to hydroxy complexes from pH > 4 and become sparingly soluble. Ascorbic acid or sulfhydryl groups of amino acids or proteins can contribute to the reduction of Fe-(III) to Fe-(II) and thus improve absorption, whereas complexing agents such as phytic acid can reduce absorption in the body (cf. Chap. 7). It is recommended that iron supplements be taken at least 1 h before meals in order to avoid complexation with food constituents. However, this then frequently leads to negative gastrointestinal symptoms. Existing anacidity of the stomach or a change in the mucosa of the small intestine in celiac disease reduce the absorption of the trace element (Farrag et al. 2019).

Following heavy supplementation with iron supplements, after saturation of transferrin, non-transferrin-bound iron (NTBI) circulates in the blood plasma where it is then weakly bound to various proteins. NTBI, even at low doses of 10 mg iron, is thought to be responsible for many known side effects such as flush syndrome and ankle edema, especially with oral iron-(II)-salts. Most iron (90%) after oral supplementation is not absorbed and can lead to severe alteration of the microbiome in the colon and inflammatory changes. Slow-release preparations can mitigate these side effects, and sustained-release preparations have comparable iron availability. Iron-(III)-polymaltose, on the other hand, leads to less formation of NTBI, which is considered very beneficial (Farrag et al. 2019; Schümann et al. 2013).

A relatively newly approved preparation is iron-(III)-maltol, which is particularly recommended for iron supplementation in chronic inflammatory bowel diseases (IBD). The complex consists of three molecules of maltol and an Fe^{3+}-ion coordinated through the three negatively charged oxygens of the hydroxyl groups. Maltol is an unsaturated cyclic ether and flavoring agent of caramel and malt, is used as a flavor enhancer in confectionery and is approved as a flavoring agent for food. Iron-(III)-maltol is reported to be well tolerated and is recommended as an oral alternative to intravenous (i.v.) iron administration in the case of mild and moderate IBD in adults.

Parenteral iron therapy directly via the blood (iron infusions) requires a prescription and can only be prescribed by a doctor. It should always be the second alternative and should only be used if there are special reasons, e.g. if the patient has not tolerated two different oral iron preparations, the oral medication is not sufficient, has an iron absorption disorder or also certain other diseases. In the past, this form of iron supplementation was not without danger, but in the meantime research has made very good progress and produced preparations, e.g. with iron carboxymaltose, which have a very high safety standard.

10.2   Food Supplements

The market research division of IQVIA (www.iqvia.com), an international group in the healthcare industry, puts the sales of dietary supplements via pharmacies in Germany at EUR 2.2 billion in 2019, with large growth rates in recent years. The market definition to the calculated figures includes oral dosage forms from the selected market segments in the field of pharmaceuticals, non-pharmaceuticals/ health products, food, cosmetics. Half of this turnover is accounted for by the group of minerals and vitamins. Of this EUR 1.1 billion, EUR 91 million is generated by the sale of iron preparations, which thus have the second strongest share among the minerals, after magnesium formulations with sales of EUR 209 million.

According to the definition, food supplements are not medicinal products, but are considered food. The Food, Commodities and Feed Code (LFGB) is responsible for this. They do not have to be approved like pharmaceuticals, but only registered with the Federal Office of Consumer Protection and Food Safety (BVL) in Braunschweig. This fundamentally distinguishes them from pharmaceuticals and should always be considered. In short, their control is by far not as strict as for pharmaceutical preparations. However, the packaging and presentation of the products often reminds one of pharmaceuticals. However, the mandatory lettering "food supplements" makes them easily recognizable as such.

Food supplements can be very valuable, but one should always make sure under which standards they have been produced and what recognized professional societies have published on the subject, especially which maximum amounts are to be observed. Under no circumstances should one try to compensate for a diet with foods of low content with food supplements (Fig. 10.1).

According to new EU law, only certain health-related claims are permitted for iron and other valuable substances. The new Food Information Regulation (LMIV) (EU 2011) introduced a mandatory nutrition declaration for food, including food

Fig. 10.1 Food supplements should not replace a balanced diet with all the necessary micro- and macronutrients. Before iron supplementation by self-medication, a physician should be consulted, who can then decide on the basis of laboratory values whether and to what extent additional iron supplementation is necessary. (© Corona Borealis/stock.adobe.com)

supplements, for the first time in Europe. Voluntary regulations were thus replaced, and the food industry now had to adapt to this new set of rules. According to the LMIV, only seven health claims relating to iron may be used on food packaging under certain conditions (EU 2012).

> **Permitted Health Claims on Iron**
> - Iron contributes to normal cognitive function
> - Iron contributes to a normal energy metabolism
> - Iron contributes to the normal formation of red blood cells and hemoglobin
> - Iron contributes to normal oxygen transport in the body
> - Iron contributes to the normal function of the immune system
> - Iron helps reduce fatigue
> - Iron has a function in cell division

These statements are then also often used on many packages of food supplements, which must also contain a corresponding minimum amount of iron. Certainly, these statements are scientifically very well verified, otherwise they would not have been included in the EU regulation. In the case of iron, however, one should be very careful because of possible harmful effects in the case of overdoses (cf. Sect. 6.3).

The Federal Institute for Risk Assessment (BfR) in Berlin states in a statement that no positive effects are known to occur with supplementation with iron in excess of requirements, while negative effects cannot be ruled out. Furthermore, the BfR recommends that only about as much iron should be taken in daily through food as corresponds to the intake recommendation, and that therefore iron should neither be used in food supplements nor in fortified foods. Iron-containing food supplements should only be taken in cases of proven iron deficiency and in consultation with a doctor (BfR 2013). A BfR publication recommends a maximum amount of 6 mg iron/day in food supplements for women aged 14–50 years. All other population groups should be warned against uncontrolled iron supplementation by a corresponding note on the products (Weißenborn et al. 2018).

Due to the particularly adverse gastrointestinal side effects of many iron supplementation preparations, there has been a great deal of research activity in the past to develop new, better tolerated formulations of iron salts.

Thus, liposomal (sucrosomal) iron is supposed to be very well bioavailable and particularly well tolerated. This is an inorganic pyrophosphate of Fe^{3+}, which is double-coated in micelles of sucrose fatty acid ester (outer layer) and phospholipid (inner layer) (Gomez-Ramirez et al. 2018) (Fig. 10.2).

A possible role of M cells is assumed for the absorption of iron-containing liposomes. These are specialized cells in the mucosal epithelium of the small intestine (Farrag et al. 2019). Liposomal formulations of active ingredients in dietary supplements are now becoming more common, and nanomedicine and nanopharmaceutics is a highly regarded research direction (Anselmo and Mitragotri 2019). When the active ingredient is linked to lipids via stronger binding forces, it is also referred to as pharmacosomes.

Furthermore, work is being done on well-tolerated oral ferritin-iron preparations from plants, especially legumes, in the USA. These are intended to be a gentler

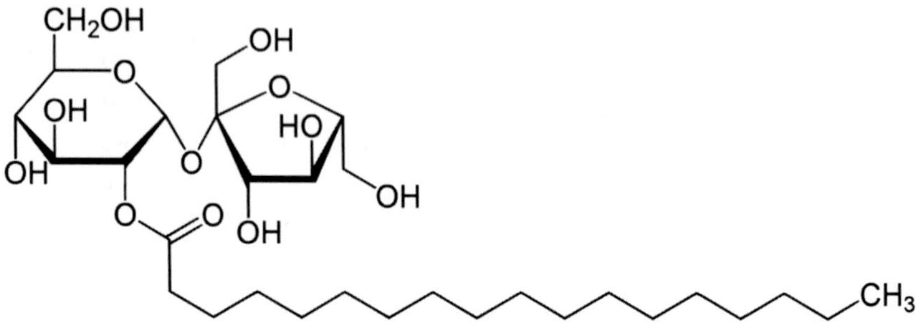

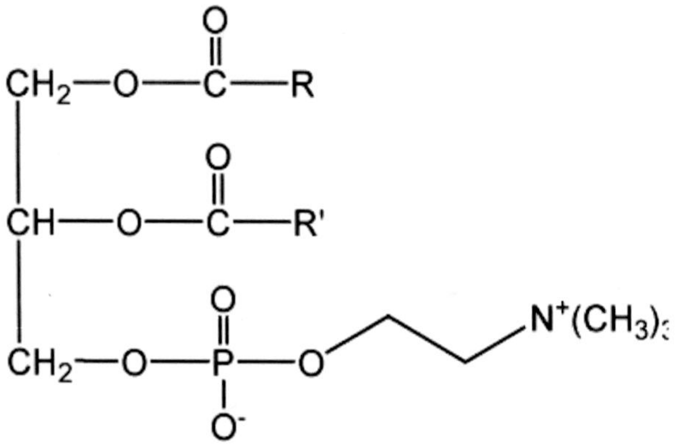

Fig. 10.2 Structural formulae of a sucrose-stearic acid ester (*sucrester*) (top) and lecithin as an example of a phospholipid (bottom)

alternative to the preparations used today, do not cause gastrointestinal problems as no Fenton-Haber-Weiss reaction takes place, and the iron is intended to have a high bioavailability (Theil 2017; Zielinska-Dawidziak 2015). In Sect. 7.3, the structure of ferritin and the newly discovered uptake pathway into the enterocytes – the ferritin port – were reported.

References

Anselmo A, Mitragotri S (2019) Nanoparticles in the clinic: an update. Bioeng Transl Med 4:e10143. https://doi.org/10.1002/btm2.10143

BfR (2013) Stellungnahme des Bundesinstituts für Risikobewertung (BfR), Nr. 016/2009 vom 2. März 2009, ergänzt am 21. Januar 2013, Verwendung von Eisen in Nahrungsergänzungsmitteln und zur Anreicherung von Lebensmitteln

EU (2011) Verordnung (EU) 1169/2011 des Europäischen Parlaments und des Rates vom 25. Oktober 2011

EU (2012) Verordnung (EU) 432/2012 der Kommission vom 16. Mai 2012

Farrag K, Lipp HP, Stein J (2019) Neue Optionen der oralen Eisentherapie. Arzneimitteltherapie 37:105–112

Gomez-Ramirez S, Brilli E, Tarantino G, Munoz M (2018) Sucrosomial iron: a new generation iron for improving oral supplementation. Pharmaceuticals 11:97. https://doi.org/10.3390/ph11040097

Roche M, Samson K, Green T, Karakochuk C, Martinez H (2021) Perspective: weekly iron and folic acid supplementation (WIFAS): a critical review and rationale for inclusion in the essential medicines list to accelerate anemia and neural tube defects reduction. Adv Nutr 12:334–342. https://doi.org/10.1093/advances/nmaa169

Schümann K, Solomons N, Orozco M, Romero-Abal M, Weiss G (2013) Differences in circulating non-transferrin-bound iron after oral administration of ferrous sulfate, sodium iron EDTA, or iron polymaltose in woman with marginal iron stores. Food Nutr Bull 34:185–193

Theil E (2017) Methods for isolation, use and analysis of ferritin. United States Patent Application Publication No.: US2017/0087209A1

Weißenborn A, Bakhiya N, Demuth I, Ehlers A, Ewald M, Niemann B, Richter K, Trefflich I, Ziegenhagen R, Hirsch-Ernst K, Lampen A (2018) Höchstmengen für Vitamine und Mineralstoffe in Nahrungsergänzungsmitteln. J Verbr Lebensm 13:25–39

Zielinska-Dawidziak M (2015) Plant ferritin – a source of iron to prevent its deficiency. Nutrients 7:1184–1201

Index

© The Author(s), under exclusive license to Springer-Verlag GmbH, DE, part of
Springer Nature 2023
K. Günther, *Diet for Iron Deficiency*,
https://doi.org/10.1007/978-3-662-65608-2

Printed in the United States
by Baker & Taylor Publisher Services